CW00500911

The Hungry Heart

Australian Inland Mission and Frontier Services 1965 – 1985

The Hungry Heart

Max Griffiths

Kangaroo Press

First published in 1992 by Kangaroo Press Pty Ltd
3 Whitehall Road (P.O. Box 75) Kenthurst 2156
Typeset by G.T. Setters Pty Limited
Printed in Singapore through Global Com Pte Ltd

ISBN 0 86417 486 1

Contents

Author's Note

Flynn of the Inland left two important legacies to the people of the Australian outback. The first was the work of the Australian Inland Mission and the second which emerged from it, was the work of the Royal Flying Doctor Service. My work in the outback stood proudly in this tradition. I owe this to Rev. Fred McKay who was John Flynn's successor as the Superintendent of the AIM and a valued colleague and friend of the Royal Flying Doctor Service.

Fred McKay encouraged my interest in the outback and invited me to join the board of the AIM. In 1965 I made an extensive visit to the remote parts of this continent, which was to be the beginning of a completely new understanding of my country. I succeeded Fred McKay as Superintendent of the AIM in 1974 and retired from the work in 1985.

I wish to acknowledge the unselfish and untiring work of the nurses, padres, teachers and others who served with the AIM and what is now called Frontier Services. The experiences I shared with them in the outback are the basis of this book.

I had the privilege of working with Aboriginal communities, mining companies and organisations such as the Isolated Childrens Parents Association as well as the Royal Flying Doctor Service. Our common endeavour was to improve the quality of life for the people who lived in Australia's isolated regions.

I am grateful for the generous support from BHP Minerals, CRA Ltd and Western Mining Corporation which enabled me to take time off to write *The Hungry Heart*. The personal encouragement of Russell Fynemore, Ray Evans and Ken Hynson was most helpful, as was the expertise of Beryl Kingdom, who overcame my inadequacies with computers.

Max Griffiths MBE
Taylors Lakes
August 1992

1

Digging Up the Desert

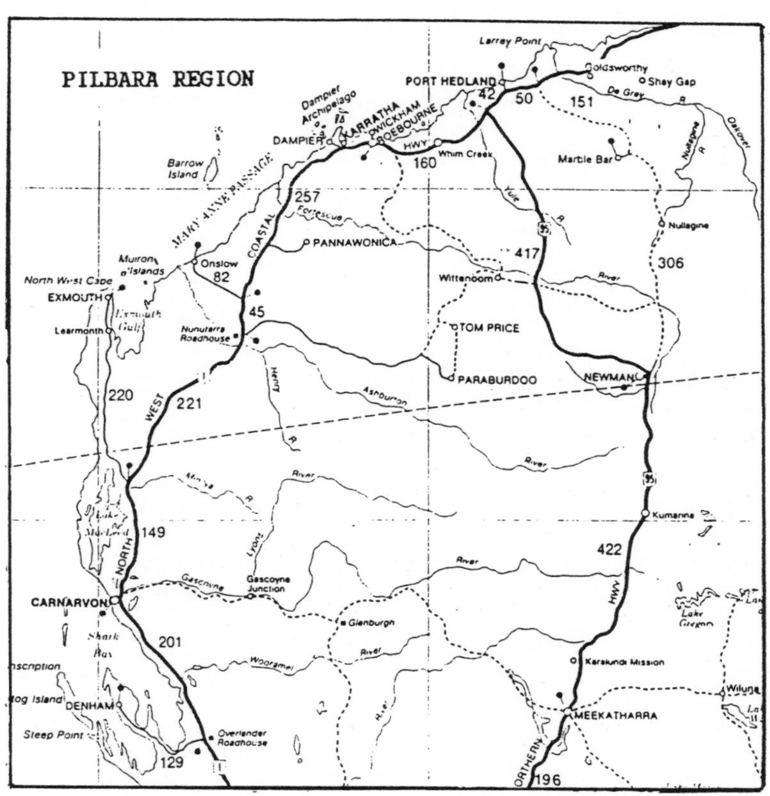

Airport terminals are not at their best in the early hours of the morning. Neither am I. The cold air clasped at my face as I stepped out of the car and hurried across to the terminal building. Inside, the atmosphere was no more welcoming. The lights were cold and yellow and my footsteps clattered along the corridors. It was in the time before the warm and cosseting comfort of departure lounges. The waiting room at Perth airport in 1965 was strictly functional.

So were the people waiting in it. They were dressed in various forms of working clothes and stood around as if waiting to be called into a courtroom to hear the verdict.

Outside, the early morning light struggled to lift the darkness of the night. The plane on which I was to travel was an old DC3. It stood with its tail on the ground and its nose and twin radial engines pointed aggressively towards the sky. The DC3 was an aerial workhorse, which had carried the human and material burdens of the world for decades. It was slow, Spartan and reliable.

A cursory voice called us to board. Reluctantly we went out into the cold morning air and shivered our way across the tarmac to the steps. Inside the darkened tunnel of the plane's cabin we found our way to the bucket seats. The plane's engines were warming up but the passengers weren't as, without any ceremony, the plane began to move slowly forward. It increased speed as it taxied down the runway, slowly turned and then proceeded to take off. After clearing the airport, the plane headed north. Our destination was the Pilbara region in the north-west of Australia.

In 1965 the north-west was the place to go in outback Australia. Vast iron ore deposits had been discovered in the Pilbara and four mining companies were beginning to dig up the desert. They were laying the foundations of what would become one of the biggest industrial enterprises Australia had ever seen.

Fifty years before, another man had made a journey north to this part of the world. His name was John Flynn and he is known to Australians as Flynn of the Inland. In 1915, Flynn travelled north in a cattle boat. His trip took a little longer than my trip in a DC3. In 1915, the north-west he visited was a wild and undeveloped frontier of Australia, almost completely isolated from the rest of the country. Fifty years later, it was still raw, undeveloped and isolated, but it would not stay that way for long.

I sat back in my seat as the hostess, wrapped in an overcoat, served me the usual plasticised breakfast. The plane lumbered along with the comfortable steadiness of its engines lulling the passengers into sleep. I wondered what John Flynn had for breakfast on the first morning of his trip and watched the endless coastline move slowly past the window. After two hours we landed at Geraldton, and two hours later at Carnarvon. We had travelled 1000 kilometres up the coast, but the real journey had not yet begun.

After Carnarvon, the plane ground steadily north for an hour or so. The coastline began to slip away to the west as we headed inland. By now the sun was high in the blue sky and the earth beneath us was turning from a dull brown to a rich red. Signs of civilisation were non-existent. The land was as empty as the sea.

Suddenly, the pilot banked steeply and the plane began to lose height. Looking out of the window I could see nothing and hoped the pilot could see more. The DC3 kept losing height until, in a final swoop, it landed with a bump. A shower of stones rattled against the fuselage. The plane came to a halt and the pilot switched off the engines. The silence in the cabin was as eerie as the emptiness outside.

'You can stretch your legs for five minutes if you like,' said the hostess.

I got out and stretched my legs. The red earth stretched to the horizon in every direction. The sun was making up for lost time and I quickly discovered that the woollen sweater I had worn in the cold atmosphere of the Perth airport was no longer necessary. The pilots stood under the shade of the wing and chatted away. They didn't seem disturbed that nothing was happening.

Then a cloud of red dust began to gather on the horizon and move towards us. Shortly afterwards, a truck pulled up. It had been a long time since it had seen a wash. A man dressed in worn khaki trousers and shirt jumped out and pulled a bag from the back of the truck. One of the pilots began pulling a few bags and packages from the baggage compartment of the plane.

'He's from a BHP exploration camp,' the hostess explained, referring to the man who had come out of the desert. 'We've brought up some spare parts they need and the week's mail.'

From the way the pilot and the BHP man exchanged good-natured banter, this was obviously a regular occurrence. The passengers slowly drifted back to the plane, the hostess slammed the door and the engines reluctantly began to turn over. As the DC3 lumbered down the strip, the man from BHP stood and watched us take off. Then he turned and walked back to his truck. The last I saw of him was a diminishing trail of dust pluming its way back to the desert.

We flew back to the coast. To my surprise, the plane continued to head out to sea. Just when I was reconciling myself to an unexpected visit to Indonesia, the engines throttled back and the plane began to lose height again. Below us was a large island. We came in to land.

Barrow Island is located about eighty kilometres off the coast of Western Australia. In 1965 it was the site of an encouraging oil discovery which was being developed by Western Australian Petroleum, or WAPET for short.

We landed on the Barrow Island airstrip, which was not much bigger than the one in the desert. A man and his truck were waiting for us. I watched a man who had disembarked feel the invisible impact of isolation in the emptiness which surrounded him. It was as if he had landed on another planet. The man with the truck exchanged more parcels and good-humoured banter with the plane's crew. Then he walked over to the new arrival and spoke to him. Obviously the new arrival had come up to join the workforce. He looked as if he had never been outside the big city before. I was to learn that, up here, he was far from alone in that respect.

We took off and headed back to the coast. Barrow Island dropped below and behind us until it disappeared into the horizon. I was learning that the new frontier of the north-west had some unexpected outposts.

In 1965, towns with airports on the coastline between Perth and Darwin were few and far between. One of them was Roebourne, a sleepy little town 200 kilometres south of Port Hedland which sprang into prominence because it had the only airport close to some of the new iron ore projects. Shortly, the town of Dampier would be built, a little to the south, with its own airport and deep-water facilities for the big ore carriers. But for the moment, the miners had to put up with Roebourne.

The plane landed on a bitumen strip. I got out of my bucket seat, said goodbye to the smiling hostess, ducked through the doorway and blinked in the bright sunshine.

'Good day. Did you bring anything for the tuckerbox?'

The lean, tanned figure in khaki who addressed me seemed no different from the mining people who were milling around. But I recognised him. He was an Australian Inland Mission padre called John. His parish was the Pilbara region and he covered a lot of ground. He had driven 600 kilometres from his home in Carnarvon to meet me. We walked over to his Land Rover, which looked as if it had survived a war, but only just.

'We'll go into Roebourne and you can buy something for the tuckerbox,' said John. 'Then we'll head for Wittenoom. I reckon we'll be on the road for two or three weeks, if you want to get a picture of what's happening up here.'

As we drove into Roebourne from the airport, I saw some Aboriginal people walking along the side of the road. They walked in single file, silhouetted against the low sky of the late afternoon. They were lean and languid. The shabby clothes they wore seemed to be an insult to what might have been a natural nakedness. Their heads were bent. It was as if their simple dignity was being crushed under the weight of an alien culture and I suddenly felt an incredible sadness.

It occurred to me that I had to travel 5000 kilometres across my country to get my first glimpse of Aboriginal people. The isolation in which they lived kept most Australians ignorant of their existence and their plight.

We filled the tuckerbox and headed for Wittenoom. The Land Rover bumped and jolted along a track that apparently everyone used, but nobody seemed to care for. After a few hours of this treatment the Land Rover decided to have a nervous breakdown. John pulled up and we got out.

'We've done a main leaf in the rear spring,' he pronounced, after inspecting the damage. 'I'll have to nurse her into Wittenoom and hope we can get a replacement.'

We arrived at Wittenoom at dusk. Fortunately there was a replacement available for the broken leaf and we spent the rest of the night replacing it. I had the feeling this was going to be an eventful trip.

'We'll head for Tom Price now,' said John the next morning. 'The construction mob have just moved in and I want to make their acquaintance.'

We travelled through some majestic country. Great cliffs of hard rock rose up before us. We plunged through gorges and crossed open plains. The Land Rover continued to bounce and buckle over the ruts in the road.

'Time to stop for lunch,' said John. 'You can light a fire and boil the billy.'

'What about the tucker?'

'I've got it under the floorboards alongside the engine. It'll be nice and warm now.'

We travelled on through the vivid country of the Pilbara. It was late in the afternoon when John told me that we must be near Tom Price. Along the road from Wittenoom we had not seen a single vehicle or person, but now I noticed a couple of caravans and a freezer truck in the scrub through which we were travelling. They were parked alongside a yard filled with cattle.

'That's the abattoirs,' John told me. 'The construction workers have to be fed, you know.'

We arrived at the Tom Price camp in the late afternoon. It consisted of a few demountable huts which served as the quarters for the men. The huts were square and squat and painted white. They stood out in strong contrast to the red, raw cliffs of the iron ore country which reared up behind them.

'What are those Toyotas doing up here?' I asked, pointing to a couple of four-wheel drives standing outside one of the huts.

'It's a sop to the Japanese,' replied John. 'We want them to buy our iron ore.'

'How do they stand up in the bush?'

'They'll never match the Land Rover,' he said with conviction. It was one of the few times I have known him to be wrong.

Then from one of the huts emerged a group of Japanese men. They were here to examine the source of their future supply of iron ore. I saw that it was not only Japanese vehicles which were invading the country. Most Australians in 1965 were blissfully unaware of the peaceful invasion which was beginning to happen in a distant region of their country.

We got out of the Land Rover and went over to the mine site office. John introduced himself to the site manager. He was astonished to discover that a padre had travelled over 1000 kilometres from Carnarvon to visit the camp. John grinned and looked at me.

'If the Board would give me an aeroplane,' he said, 'I would be up here more often.'

'Well,' said the manager, 'any time you want to drop in, there will be a bed and a meal for you.'

We stayed the night in the camp and ate in the mess with the men. Afterwards, we sat around and yarned. The men seemed pleased to see us. Some of them were still recovering from the shock of being in this isolated spot in the desert.

'One day I was happily working in my office in Melbourne,' said one man, 'going home to my pleasant suburb at night. I used to do some opera singing in my spare time.' Then he blinked as if he still couldn't believe it. 'The next day I was packing my bags to come up here.'

He looked around the mess, with its bare walls and basic furniture.

'My family has got as far as Perth. When we get the town built, they'll come up and join me.' Then he turned and grinned at me. 'That's the mining industry. You're either on top of the world or at the bottom of the pit.'

I was to hear similar stories many times after, from people whose lives had undergone radical change as a result of coming to the Pilbara. Some made the adjustment remarkably well. For others, it was a painful and short-lived experience. But you can't set out to dig up the desert on the scale that was happening in the Pilbara and not disrupt a few lives.

The first arrivals were the construction workers. Most of them were old hands at moving and their family life was accustomed to coping with it. Some came for two weeks and others for two years, depending on the nature of their work. The construction phase of any new project is intense and often punctuated with catastrophes. In an isolated region such as the Pilbara, the catastrophes were more frequent and on a grander scale.

We saw one dramatic example of this in the course of travelling around the Pilbara. As we drove towards King Bay on the way back from the interior, the burnt orange of the land came into colourful confrontation with the blue of the glittering sea. A cluster of low buildings on the horizon marked the construction site for the new port of Dampier. One of the buildings, longer than the others, had been wrecked. It looked as if a giant hand had smashed down on it.

'It's the mess building,' explained a worker, when we stopped to ask. 'A grader driver did it. The pressure got to him and he went berserk. He wiped out two-thirds of the mess. We put him on the next plane back to Perth, but the men were bloody annoyed. The mess is all they've got up here apart from the work.'

Planes flew in and out every day, as the construction work increased in tempo. They brought in new workers and took others out. A large number of those who left simply couldn't take the pressure. Others seemed to thrive on it. Some of the men who pioneered the Pilbara ended up on the top of the corporate tree in the mining industry. Others disappeared into oblivion.

What I saw in the Pilbara in 1965 was the beginning of a modern miracle. Despite its extreme isolation and the shortage of people with the skills and experience to undertake such a giant project, the iron ore industry was established in a miraculously short time.

Take Hamersley Iron Company, for example. They had to build a port capable of accommodating ships bigger than any that had ever come to Australia. The mine site at Tom Price was 500 kilometres inland. Between the mine and the port, a railway line was laid which had to carry ore trains a mile long, with loads of over 20 000 tons.

The whole project consisted of two towns, a port, a mine and 500 kilometres of railway line. It was finished in nineteen months. Twenty years later, one of the Hamersley executives said to me, 'If we had been trying to do that today, we wouldn't have finished the environmental impact study in nineteen months'.

Travelling through the Pilbara in 1965 was like watching history being written with a large bulldozer. But my reasons for being in the Pilbara were more than to observe history in the making.

John Flynn came to the Pilbara fifty years earlier because of his concern for people living in isolation. In 1965, isolation was still to be the fate for thousands of people who came to take up residence in the new mining towns. They were a little different from the rough shacks which housed the miners on the gold diggings, but the Pilbara was still the place where isolation could grip people in its fearful hands. If we were going to continue the Flynn tradition, John the padre would need something better than a battered old Land Rover to travel the 1000 kilometres from Carnarvon to Tom Price. He would need a plane.

We left the camp site at Tom Price and travelled south, deep into the ranges of the Ashburton River country. John had already visited this part of the country and was going to call in at a station property where he had been made welcome on a previous occasion. We pulled up at a gate and I hopped out to open it. As I climbed back into the Land Rover, John looked at me and said:

'This woman we are going to meet. If you pull out your camera and start taking photographs, I'll break your bloody neck.'

I carefully stowed my camera under the seat.

The woman stood on the homestead verandah, waiting for us. She had the look of someone who had gone to the letter box and found nothing. We were invited in. The house had the same atmosphere of life without expectation. The floor was the simple bare earth, beaten into submission. We sat down on an old lounge suite and tried to make conversation. It did not come easily. The woman was not impolite. Isolation had robbed her of the art of conversation.

'Ted's at the stock camp,' she said, explaining her husband's absence. 'He'll be away for the next six weeks.'

John asked if she had seen any of her neighbours lately and she said no. He asked if she and her husband would be going to the race meeting next month. The races were held once a year in Onslow, a small town on the coast. The woman said she didn't think so as her husband was busy.

'But all your friends will be there,' said John. 'They'll be disappointed if you don't go. Besides,' he added with a laugh, 'it's a good excuse to buy a new hat.'

The woman smiled thinly.

'I'll think about it,' she said.

I began to wonder how many other women she had spoken to over the last twelve months. Perhaps one or two; maybe none. I looked out the window, across the empty spaces that stretched to the horizon and the empty sky that stretched beyond it. Onslow seemed a long way away.

We left the homestead and headed back down the track to the station gate. I looked back at the woman standing on the homestead verandah and remembered the man standing on the airstrip at Barrow Island. I was beginning to understand the meaning of isolation.

Within a short time I had seen the old and the new in the Pilbara. On the one hand there were the miners and their families, who would come in their thousands and bring a whole new way of life to the region. On the other hand there were a few people like the woman at the homestead who were living a losing battle against nature's ruthless judgment and feeling the weight of its sentence.

Yet people stayed and survived. I wondered what their secret was. At night in the loneliness of the bush, the silent wings of God brushed my soul and caused it to tremble. I wondered if I would ever learn how people coped with living in isolation. As it happened, I stumbled on part of the secret quite by accident.

Leaving the homestead and the Ashburton country, we headed back to the coast. We called in at the town of Onslow and then drove up the north-west cape to the site of the new United States naval wireless base at Exmouth. Then we turned and headed south to Carnarvon.

Despite the absorbing interest of the Pilbara developments, we were looking forward to a bath and a bed. It had been a long trip and the tracks had jolted and jarred our bones. We would be glad to wash the dust of the Pilbara out of our clothes. Fortunately the weather had been kind the whole way and we had avoided the frustrations and delays of being bogged.

Nature, however, tends to demand the last laugh. On the track down from Exmouth the heavens suddenly opened up and the rains came tumbling down. We began the race to get back before the roads flooded and nature gave us one last lesson in isolation. John pushed the Land Rover's engine to full capacity. The engine roared in protest. But we made it. When you are as close as that, desperation is a powerful force to drive you over the line.

The driveway into John's home in Carnarvon was deep in the shadows of the night. We stepped out knee deep into water. Two dirty and dishevelled figures approached the front door, which was opened by John's wife, Olive.

'Don't bother to come in,' was her greeting. 'Someone urgently needs your help. They're at the motel down the street.'

We turned back into the night and drove to the motel.

'There's a couple in Unit 12,' the manager told us. 'They asked for a minister.'

We knocked on the door of Unit 12, conscious of the fact that the last thing we looked like was a couple of respectable parsons. The door was opened by a pleasant middle-aged man who asked us in and introduced us to his wife. They were country people by the look of them.

'It's our daughter,' said the man. 'She's getting married tomorrow up at Onslow. Do you know it?'

We assured him that we had just come from Onslow and had heard about the wedding.

'Well, the Anglican minister here was to have conducted the service. But his father died in Perth this morning and he had to go down and make the arrangements. They need someone to conduct the service and there doesn't seem to be another minister for a thousand miles.' Then he looked even more troubled. 'The other problem is that we have driven up from our wheat farm in the south and now we hear that the road north is flooded and we can't get through.'

We knew all about that too. John thought for a moment.

'The wedding's not a problem,' he said cheerfully. 'My friend here is a minister and when we scrub him up a bit he'll be quite presentable.'

The coupled looked at me doubtfully and, catching a glimpse of myself in the mirror, I didn't blame them.

'Now regarding getting there,' John went on, 'I have a mate in town who flies a plane. I'll see if he is free to fly you.'

John rang his friend and returned with the news that the plane would be available. We would fly to Onslow in the morning, have the wedding in the afternoon and return the next day. The couple were greatly relieved. There were a few other problems, such as the fact that all my 'how to conduct a wedding' equipment was on the other side of the continent. But by this time I was fired up with all the enterprise and initiative I had seen in the Pilbara. I could have conducted a wedding on the moon.

By the morning the rain had cleared and we were able to fly the 400 kilometres to Onslow in cloudless skies. The road below us looked like a river that had broken its banks. We were all looking down on the countryside and sitting in the silence of our thoughts when the mother of the bride let out an anguished cry.

'What is it?' asked her husband anxiously.

'My hat! I've left my hat behind,' she wailed.

We sympathised at the prospect of the mother of the bride without a hat but the pilot, having flown a couple of hundred kilometres, showed no inclination to turn around and go back.

We landed at Onslow and were driven into town. The bride was a teacher at the local school and the groom was the district representative of a big pastoral company. They were a young, pleasant couple who had quickly adapted to the changed circumstances.

'This will be the first wedding to be held in Onslow for twenty years,' said the bridegroom. 'We are about to make history.'

Just how much history I was about to learn.

The bride took me to one side. 'I hope you don't mind,' she said, 'but the children at my school asked if they could do something for the wedding. I've let them decorate the church. Most of them are Aborigines and have never seen a wedding before.'

I saw no problem.

'Better go and have a look,' said the bride.

I went down to the church. It was a tiny weatherboard building. The paint had been peeled off by the ravages of the weather. There was a cross fixed to the roof and it was slightly askew. I went inside.

The church was decorated with balloons and streamers and looked as if the scene had been set for a great party. I thought of all the austere Gothic cathedral type churches in which I had conducted weddings. Now I was certain we were about to make history.

Outback weddings generally follow a tradition. There is the service in the church or under a tree. Then the family and a few friends go to the local hotel for a meal. The big celebration is held later in the night. All it needs is a barn, a bar and a band. They seem to materialise, even in the most remote places.

The celebrations after the wedding at Onslow went with a swing. Because the couple were so well known, everybody in the region had been invited. Judging by the noise and the crush in the hall, everyone had come. One of the men behind the bar asked me if I would take over for a while. He wanted to have a dance with the bride. In view of the fact that another twenty years might pass before he got another chance, I was happy to oblige.

I was introduced to some of the station people in the region and invited to drop in for breakfast the next morning. Some of the stations were hundreds of kilometres away, but that didn't seem to matter.

During an unexpected lull in the bar trade I looked up and saw a group of men who had come into the hall and were watching the dancing. They were dressed in the khaki uniform of miners and looked as if they had just come out of the desert. I recognised one of them and caught his eye. He came over for a drink.

We had been fellow students at Melbourne University and he was now a geologist with BHP. His group had been doing some survey work in the region and their guide, who lived in Onslow, had invited them to the wedding. We joked about the difference between the formal dining halls of the university colleges in which we had once lived and the riotous atmosphere of the dim, dusty barn in which we now stood.

I left the celebration some hours later and tried to get some sleep before flying back to Carnarvon in the morning. After a snatch of sleep, we took off and headed south.

As we flew back, I had time to reflect on my experiences. I thought I had glimpsed something of the secret of survival in the outback. Coping with isolation means finding opportunities to meet and mix with other people. The opportunities do not come often and sometimes, as with the woman on the lonely pastoral station, it may be once a year, at a race meeting. Sometimes, as with the wedding in the remote little town of Onslow, it might happen only once in twenty years.

It was not a hell of a lot of opportunity, but you made the most of it.

I sat back in the plane and thought again about the woman in that lonely homestead. I hoped she went to the races this year.

2

Storm Clouds over the Kimberley

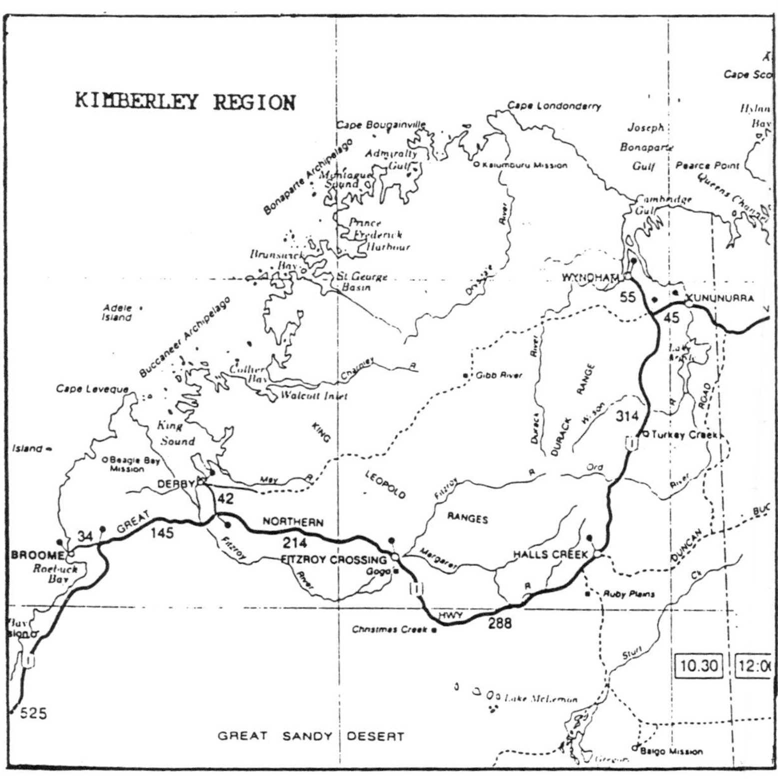

‘Well, the north-west is certainly the place to go.’

The man standing next to me in the Port Hedland airport terminal smiled appreciatively.

‘Where have you been?’ he enquired.

‘I’ve just had a look at some of the new mining ventures here in the Pilbara,’ I replied.

The Pilbara, 1965. An inhospitable country waiting to be turned upside down by the iron ore mining industry.

Mechanical monsters in the desert. A drag line scoops iron ore from the mine at Tom Price in the Pilbara.

Where Aboriginal health problems were first tackled seriously. Fitzroy Crossing Hospital built by Flynn in 1939.

River crossing in the Kimberley. In the wet seasons the nurses took to the boats.

'Well, you've only seen half of it,' said the man, with some authority. 'Wait till you've seen the Kimberley.'

Roy Hamilton was the Director of the Office of the North-West for the Western Australian government. The opening up of the Pilbara was only part of his job. The government also had ambitious plans for the Kimberley region, which was part of the north-west.

'I'm heading for the Kimberley now,' I explained to Roy. 'What's it like?'

'Some people see it as the most turbulent, soul-destroying region in Australia. Others reckon it's the most awe-inspiring and beautiful place on earth. It all depends on when you visit.' Then Roy grinned. 'You're lucky. You're going in the best time of the year.'

Roy went on to tell me of the plans the government had for the development of the Kimberley:

'We're going to build a vast irrigation scheme by damming the Ord River. Kimberley country is tropical country. In the wet season the rain fills the rivers and floods the countryside. By damming the Ord we'll create a water reservoir seven times the size of Sydney Harbour.'

The plan was to grow cotton on the irrigation farms and sell the crop to Asia. Politicians were already talking about the Kimberley region as the 'clothes basket of Asia'.

Tragically, the Ord River Scheme, as it was called, became the dream that drowned. Disease struck the cotton and wiped out the crops. Today, the dam is called Lake Argyle and is a pleasant tourist attraction. But in 1965, as I talked to Roy Hamilton in the Port Hedland airport terminal, the dreams still dominated and if you made a trip to the north-west to see the Pilbara iron ore projects, it was mandatory to go on and see the Ord River Scheme in the Kimberley.

The Ord River Scheme wasn't the only reason I was flying towards the Kimberley. When Flynn of the Inland travelled to the Pilbara in 1915 he also went on in his cattle boat to the Kimberley region. At that time, Broome was the centre for a large pearlling industry. Hundreds of young men worked there. But outside their work they had nothing to do. Flynn established a welfare club in Broome, where the young men came to read, play games and talk.

The Kimberley region was also the place where John Flynn first saw the appalling conditions in which Aboriginal people lived. He used one word to describe these conditions—'Rotten!'

Fifty years later, when I saw the conditions in which the Aborigines lived, I had no reason to modify Flynn's comments. But on the day I flew from the Pilbara to the Kimberley the weather seduced me into believing that it was the most admirable of all regions in Australia.

The old DC3 in which we flew carried a mixture of passengers and cargo, lining either side of the cabin. All the passengers, with the exception of me, were headed to the races at Broome. Some of them looked as if they would go to the races in Hell, if the odds were right.

We let down at Broome and I wished them all luck, bookmakers included. Then the plane took off for the final leg to Derby. Having nothing else to do, I surveyed the cargo. Amongst it, and enticingly close, was a bicycle. I persuaded the hostess that the pilots needed her attention much more than I did and having got rid of her proceeded to become the first person to ride a bike at 12 000 feet.

Derby is one of the two Kimberley ports for the export of cattle. The other is Wyndham. Cynics will tell you that if the cattle feel they have to leave Derby and Wyndham, there is no excuse for humans to stay. We landed at Derby. There was no crowd waiting to greet the plane and only one man who came out to collect the baggage and freight. I went and asked him for my bags. He said there was no baggage marked for Derby. Since there were no other passengers on the plane, I suggested that it was logical that any baggage in the hold might be mine. Logic has no persuasive power in the Kimberley, but the charm of a hostess does. I got my bag.

A lean khaki-clad figure detached itself from the airport fence and walked towards me.

'Now that we've got all that sorted out,' he said, 'welcome to the Kimberley.'

It was the patrol padre who worked in the Kimberley. His name was David and he had travelled 1000 kilometres to meet me, further even than John the padre in the Pilbara. I picked up my swag and threw it into the back of his truck. We left Derby and headed inland.

As we travelled, David told me something of the Kimberley region. Apparently life there is never dull and if all else failed, the weather could be relied on to create a diversion.

The wet season was destructive to people and property alike. Flooded rivers drowned thousands of head of cattle and put roads out of action for months on end. Temperatures soared to 45°C and stayed there. The humidity drained people of their energy and left their passions raw. Tempers flared between the best of friends. Others sucked their feelings inwards and became suicidal. Human relations got the thick end of the stick.

But the aching tooth of the Kimberley was the Aboriginal situation. In Flynn's word, it was rotten and like an aching tooth it would be painful to remove. David said that when we reached Fitzroy Crossing I would see what he meant.

After leaving the coast, we passed through some flat and uninteresting country. The only attraction to my curiosity was the bottle-shaped boab tree, which cropped up intermittently along the road. Further inland the country began to develop character. The sweeping plains were brought to an abrupt halt by chains of rugged ranges that ran along the horizons. The countryside began to develop colour, as sunshine and cloud painted the ranges green and grey and blue. The Kimberley was absorbing me into its ambience.

We came out onto a clear plain. David pointed to the horizon ahead of us.

'This is Plum Plain,' he said, 'and Fitzroy Crossing is on the other side.'

The mouth of the Fitzroy River is just south of Derby. Unknown to me, we had been following a track parallel to the river for about 300 kilometres. In front of us, the river turned sharply north and came into confrontation with the road along which we were travelling. Hence the name, Fitzroy Crossing. In 1939, John Flynn had built a hospital there and the Australian Inland Mission still conducted it. It was my first visit and I was looking forward to seeing this legendary medical outpost.

We came to a creek and David negotiated the causeway. There was nothing in sight. A little later, he turned off the road and headed towards a two-storey building, partly concealed among the trees. We pulled up and got out.

'Here we are,' announced David. 'Welcome to the Fitzroy Crossing hospital.'

I stood in silence.

'What do you think of it?' he asked.

I fumbled for the appropriate word. After a few minutes it came to me. 'Dilapidated.'

'Wait till you see the inside,' he replied.

The building was clad with sheets of corrugated iron. Large sections of the front wall opened outwards to become awnings. Attempts to cultivate little pieces of lawn and a garden had failed. The flywire on the front door had come away from the frame. The weather had flayed the paintwork.

David was right. The inside was no improvement. A long verandah with wire screens ran along the front of the building. There were beds which looked as if Florence Nightingale might have used them in the Crimean War. Aboriginal patients lay on the beds or sat on the concrete floor.

We went through to a large room which seemed to be a combined kitchen, dining and public space. There were two women dressed in the traditional white dress of nurses. We were introduced. After a cup of tea, one of the nurses took me on a tour of inspection.

The room used as a clinic was small, gloomy and lacked equipment. The patients' toilet and bathroom area was drab and needed a coat of paint. Then the nurse took me upstairs.

'Be careful,' she said. 'The white ants have been at work.'

I stared in amazement. Whole sections of the floorboards had been removed and some of the bearers showed signs of having been eaten away. Getting across to the other side of the top floor would require a tightrope performance, with the possibility of the bearers collapsing underneath you.

'What happens up here?' I asked.

'It's supposed to be out of bounds, but we need the space. One of us has to sleep up here, because there's not enough room downstairs. It's a bit tricky when you have to get up in the middle of the night with a torch, to go and attend to a patient.'

We went back downstairs.

'Which is the other nurse's bedroom?' I asked.

The nurse opened the door of a room and we walked in. Lined up around the walls of the bedroom were a number of old wooden boxes. They were all filled with Aboriginal babies.

'We had a gastro epidemic,' explained the nurse. 'I put them in here because there was nowhere else. In any case I can keep an eye on them during the night.'

'Do you ever get any sleep up here?' I enquired.

We went out the back door of the hospital and into the yard. I blinked in the strong sunlight. The nurse pointed to a broken-down shed at the back of the yard.

'That's the native ward,' she said. 'Of course, we don't use it now.'

I looked at the shed which had the appearance of a dilapidated fowl house and wondered how it ever could have been used for human beings, let alone sick ones. The nurse saw the look on my face and began to explain that in past times it had been necessary to isolate Aborigines who had serious diseases. Of course the Aborigines felt more comfortable in the open.

Of course!

The sun beat down and I closed my eyes to shut out the glare. But I couldn't shut out the sight of what I had seen. I told myself that this was twentieth-century Australia and the second half of the century at that. But the Australia I knew seemed a million miles away.

We walked across to another corrugated-iron shed with a windmill alongside it. There were a number of forty-four-gallon drums against the outside wall.

'This is the engine shed,' said the nurse. 'We generate our own electricity and pump our water with the windmill and an auxiliary engine.'

'Who looks after the engines?' I asked.

'We do. I knew nothing about engines before I came to Fitzroy Crossing. I had to learn fast.'

I went inside the shed. It was dark and smelled strongly of oil. I peered at the engines, which appeared to be rather elderly, and asked if they broke down often.

For the first time the nurse's face lost its look of composure.

'Too often,' she said curtly. 'And it's no fun being in the middle of a difficult stitching job in the clinic when the lights go out.'

Suddenly the sun seemed unbearably hot, so we returned into the cool interior of the hospital, where David greeted me.

'Come on,' he said. 'The nurses have better things to do than entertain a couple of visitors. I'll take you on a tour of Fitzroy Crossing.'

The hospital was built close to the bank of the Fitzroy River. We walked across and looked down the steep incline to the river bed, about twelve metres below. The road along which we had driven went straight down the bank and became a concrete causeway across the dry bed. Then it made an equally steep climb up the other side. While we were looking down, I heard the sound of a large truck approaching.

'That's a cattle train,' said David. 'The drivers have some fun crossing at this place.'

The cattle train was a truck with a prime mover towing three large trailers, each of which was filled with cattle. The driver stopped at the edge of the river bank, changed into low gear and carefully descended to the bottom. He crossed the causeway and the truck groaned and grunted its way up the other side. So did the cattle it was carrying.

'What happens in the wet season?' I asked.

'The water rises right up to the top of the banks and generally floods over the surrounding country,' said David. 'It's impossible to get across for weeks at a time. That means the hospital is cut off from the rest of the world.'

I made one of the many extraordinary mental notes that I compiled on the trip. This one was to persuade someone to buy a boat for the Fitzroy Crossing hospital.

'Now show me the town,' I said to David.

'Well, there isn't one to show,' he replied.

We drove back from the river, past the hospital. David pointed out two other buildings which were set back among the trees. They were a short distance from the hospital and from each other. The first was a post office and the second was a police station. They were made from timber, painted green and raised from the ground to protect them from flood waters. We crossed a creek and turned left into the bush country on the other side. Then we came to a large, sprawling, unimpressive building. A couple of begrimed bush vehicles stood outside.

'The Crossing Inn, the local pub,' David explained. 'The part that has been added to the end is the store.'

I drank in the atmosphere of this thriving hub of the community. Then we turned around and drove for a couple of kilometres in another direction until we came to a compound of old bush buildings. There were a number of Aborigines drifting about.

'This is the mission,' David told me. 'There are about 200 Aborigines here. They have their own school.' Then he turned the vehicle around and said, 'And now you have seen Fitzroy Crossing.'

A hospital, a post office, a police station and a pub, together with a mission full of Aborigines, as far apart from each other as you can get without moving into the next region.

'Don't the people up here speak to each other?' I asked.

'You don't know how right you are,' he replied, laughing.

Later in the day, I borrowed the vehicle and went back to the pub, determined to discover something more about this extraordinary place. A pub in the bush is always a good place to paint some colour into the local scene.

I pulled up outside the Crossing Inn and went into the bar. The area was roofed, but otherwise open to the world in every direction. The serving counter split the bar area down the centre. In later times, it was the dividing line between Aboriginal and white drinkers. I ordered a drink and stood back to observe.

There was one other customer in the bar. Obviously a stockman, he stood at the bar with one boot hooked over the rail. His cattleman's hat was pulled down and I could not see his face. We stood for a while, gazing across the open space into the separate projection of our thoughts. Then he spoke to me.

'What are you doing up here?'

It was spoken by way of a challenge and I wasn't sure how to answer it.

'I'm up visiting the hospital,' I finally said.

He thought about that for a while. 'You work them like bloody horses,' he said.

It was my turn to be puzzled. After a while I realised he was talking about the nurses.

'Yes, I know,' I replied. I made some lame excuse about the government not giving us any money to do the job, but it sounded inadequate.

Then he turned and looked at me for the first time.

'Well, if you haven't got any money, I suppose I'd better buy you a beer.'

He told me he worked on one of the cattle stations in the district and had come in to pick up the mail, buy a few provisions and drop one of the Aborigines off at the hospital for treatment. He'd be going back to pick up the Aborigine in a few minutes and drop off a sugar bag of meat for the nurses. I asked him whether he often brought in sick Aborigines for treatment.

'Sometimes one or two,' he replied. 'Sometimes a whole truck load.'

I thought about the roomful of sick babies in boxes at the hospital and wondered if they had travelled from the station in the back of a truck.

'I'll have a look at the engine while I'm there,' the stockman went on. 'It's about time you replaced it.'

I made another entry in my mental list.

The Fitzroy Crossing hospital which John Flynn had built in 1939 was designed to withstand the ferocity of sun and storm. But time is always on nature's side. The nearest tradespeople who could service the engines or repair the building were 300 kilometres away in Derby and were never enthusiastic about making the trip. The sewerage system broke down frequently and created serious problems of infection. Repairs or replacements of medical equipment had to be ordered from Perth, 3000 kilometres away.

'Sometimes,' a nurse said to me, 'you'd think they were still using the cattle boat in which John Flynn travelled.'

The accumulation of repairs needed at the hospital had almost reached the point of no return. So had the nurses. But what was causing them even greater concern was the deteriorating social scene in Fitzroy Crossing.

Up until the mid-1960s most of the Aborigines lived on cattle stations, in camps of up to 300. Their only contact with the outside world was when they came into the hospital for treatment. Around 1965 there was the beginning of a movement which saw the Aborigines move away from the stations. This would gather momentum a few years later, when Aboriginal people were granted full citizenship rights, but in 1965 the movement had already started. A number of Aborigines had come from the stations and settled in camps around Fitzroy Crossing itself.

The hospital, already under pressure because of its deteriorating condition, came under further pressure from the increase in Aboriginal patients. I stood in the hospital clinic one afternoon during my visit and watched one of the nurses treating Aboriginal patients.

'Well, you may as well make yourself useful,' she said to me after a while. 'Hold this woman's hand for me while I get a dressing.'

I looked at the woman's hand. She had a condition I could not identify. When the nurse returned, I asked her what it was.

'Oh,' she said casually, 'that's leprosy.'

My mind did a somersault back to university days. A friend who was doing final-year medicine was well in the running for a prestigious prize. He was presented with a patient and asked to make a diagnosis. After a while, he said to himself that if it weren't the twentieth century and he weren't in Melbourne, he'd swear it was leprosy. But he dismissed the idea as ridiculous. I now knew how he felt when they told him it was.

The nurses told me about the appalling infant mortality rate among the Aborigines. They described their frustrations when patients were brought in from the stations, almost too late for them to do anything. Their difficulties were aggravated by the fact that, in cases of emergency, it was necessary to call the flying doctor plane to carry out an evacuation to Derby. That might takes hours to accomplish.

I looked out at the brilliant sunshine which was beating down on the Kimberley countryside. Away to the horizon, I could see storm clouds beginning to gather. There were other clouds gathering over the Kimberley in the form of the Aboriginal situation. I wondered when the storm would break.

Feeling a little guilty about the inadequacy of the hospital, I asked if there was anything I could do.

'The patients' toilet is out of action.' said one of the nurses. 'You can fix that.'

David had gone on a visit to one of the stations, so I turned off the water main and went to work. The problem seemed to be a blockage in one of the pipes. I unscrewed it and pulled away at the offending blockage. It proved to be stubborn.

In the meantime, David had returned from his visit and finding the water turned off, cursed the fool who had done it. He proceeded to turn it on again. I was just in the act of removing the blockage when the water came rushing down the pipe. My baptism into Aboriginal affairs was rather dramatic in more senses than one.

'You won't understand the Kimberley until you've been on a cattle station. I have to call into one of them on the way out.'

We had left Fitzroy Crossing and were heading inland. David spent a lot of time on the road, visiting the many cattle stations scattered across the Kimberley region. We left the main road and headed down a bush track.

David parked his truck under the shade of a tree outside the homestead. It was a large building with wide verandahs running around it. The woman who met us was the owner's wife. She was fair complexioned and wore a simple cotton frock with quiet elegance. We went into a large sitting room with comfortable chairs and wide windows.

I discovered that we came from the same part of Australia and there were a few people whom we both knew. We laughed at the contrast between the emerald green sheep country from which she had come and the bare brown cattle country in which she now lived.

I asked about the Aborigines on the station. We had passed a large camp of them close to the track on the way in. A frown cast a shadow over the woman's face.

'They live in their camp and we don't see a lot of them,' she said, 'except the few who work around the place.' Then she pointed to a small building not far from the homestead. 'That's the school. The children go there. What happens after that—'

She shrugged her shoulders.

We went across to the school and met the young teacher. He was enthusiastic and the children seemed happy. But then as I discovered later, Aboriginal children generally do.

Back at the homestead, I asked the woman how she coped with the wet season.

'We go south,' she said promptly. 'Most of the white women do. I know it sounds like running away, but the truth is, that's the only way we preserve our sanity.'

As we waved goodbye and our truck churned the dust down the track away from the homestead, I looked back at a solitary figure who stood at the gate watching us. There was not one trace of unhappiness or anxiety in her bearing. But I thought I knew where her heart was.

The road inland from Fitzroy Crossing leads to Halls Creek, about 300 kilometres. In 1965, Halls Creek was to Fitzroy Crossing as Melbourne is to Sydney. The best thing you can say to the locals is that the two places are, well, different.

Halls Creek had streets and shops and a town power station. The nurses in the hospital had been liberated from the slavery of attending to crotchety old diesel engines. They could walk out of the hospital, cross the street and stroll down a footpath to the shops. They greeted passers by, including the Aboriginal people. Halls Creek was a town and a community, where Fitzroy Crossing was a handful of scattered buildings.

The Halls Creek hospital was more functional than Fitzroy Crossing. But the health problems of the Aboriginal people were no different. Station trucks pulled up and unloaded the sick, then came and picked them up again when the station worker had finished his other business. The most common complaints were eye, ear, chest and stomach infections. There were also infected wounds that had been left too long before they were brought into the hospital for treatment. The nurses expressed their frustration to me.

'You almost give up,' said one of them. 'They come in far too late. You treat them, knowing that the same thing will happen again despite what you say about not leaving it so long next time. It's no wonder that so many of the children die.'

I asked one of the nurses what she thought of the Aborigines.

'When I first came up here I was appalled by the conditions in which they lived. To be honest, I couldn't cope. I found myself getting angry because they didn't seem to want to do anything about it. If I lectured the mothers about looking after their children, they simply smiled and said, "Yes, sister". But next week they would be back with the same condition.'

'How do you feel now?'

She thought for a moment.

'It's different. Once I got over the initial shock I found them to be a very gentle people. They really care for their children. But their ways are not our ways. Some of the women take me out in the bush occasionally.' She laughed. 'I guess it's because I've got a vehicle. But we have a lot of fun. They're quite different out there. They poke fun at me, in a gentle way. We catch fish and make damper. Out there I can understand what their life is all about. In here I can only see the futility of it all.'

I was beginning to see that it would take more than a new hospital at Fitzroy Crossing to solve the Aboriginal health problem.

As I had gone to the Fitzroy Crossing pub to take in the local colour, so I wondered where to go in Halls Creek. In the course of my travels I had ripped a pair of trousers and needed a new pair.

'Go and see Mrs Johnston,' said the nurses. 'She'll fix you up.'

I stepped out into the sunshine and wandered down to the centre of town. It was not hard to find Johnston's General Store. I walked in the open door. Mrs Johnston sat on a stool behind the counter, guarding the till. She eyed me with suspicion.

'The nurses said I could buy a pair of trousers here,' I said hopefully.

Mrs Johnston eyed me up and down as if measuring me for a coffin.

'Go in there,' she ordered, pointing to a doorway with a beaded curtain. I went in and found myself sharing a small room with several sacks of onions and potatoes. After a while Mrs Johnston came in with a pair of trousers in one hand.

'Take those off,' she commanded, pointing to the pair I was wearing. Meekly I obeyed. 'Now try these on.'

Mrs Johnston thrust the pair she was holding into my hands and watched as I hastily pulled on the new pair of pants. She surveyed me for a moment.

'They'll do,' she decided and turned and walked back to the shop.

I had a wonderful vision of Mrs Johnston in charge of the Henry Buck's exclusive menswear store in Melbourne. I kept going back to see Mrs Johnston on subsequent visits to Halls Creek, but I bought my trousers elsewhere.

The main road leaves Halls Creek, turns sharply north and heads towards Kununurra and Wyndham. It is a road worth travelling. There are ranges running parallel to the road that create a purple and blue backdrop to the green and brown of the plains. In the late afternoon it is a deeply satisfying sight.

'We'll turn off and head for Springvale station,' said David. 'They're expecting us to stay the night.'

Springvale station was the home of a legendary Kimberley character called Tom Quilty. Tom owned several stations in the region and a share in the meatworks at Wyndham. Surprisingly, he also wrote poetry. It was an earthy kind which would not win him a Nobel Prize for literature, but the bush people loved it. The homestead was a rambling old building with a central courtyard area.

We arrived in time for dinner. I noted that there were a number of Aboriginal people around and the children seemed to be quite at home.

'Most of them are called Quilty,' said David.

After dinner Tom's wife, Olive, excused herself and Tom led us into his study. It was filled with overstuffed armchairs. The walls were covered with cattlemen trophies. In one corner, there was a bar.

'We drink rum here,' said Tom. 'Help yourself.'

We sat down and Tom started to ask a lot of provocative questions. When I tried to answer them, he proceeded to argue with my statements. It would have been easy to dismiss him as an arrogant prejudiced cattleman, but I suspected there was a shrewd brain beneath this aggressiveness. He was testing me, to see how I stood up to this kind of treatment.

'I had a welfare fellow up here the other day,' said Tom. 'Wanted me to give the Aborigines orange juice. Orange juice! I can't afford to buy oranges for myself, even if I could get them.'

I explained that unless the Aborigines received some form of vitamin C, their health problems would never be solved.

'Rubbish! They eat their own bush tucker and we give them all the meat and bread they want. Feed them well, too!'

I could see the welfare man had a long road ahead of him.

We talked far into the night. Tom kept on insisting that I fill my glass from the rum bottle. After a while, I became desperate. Then I noticed a bottle of dry ginger ale on top of the bar. Placing my back between Tom and the bottle, I was able to continue the rest of the night drinking ginger ale. We finally got to bed.

Next morning, as we were leaving, Tom drew me aside and slipped a cheque into my hand. It was a generous donation for the AIM.

'You drank me under the table last night,' he acknowledged. 'It's never happened before.'

We were now on the final stage of our journey through the Kimberley. We left Springvale and drove towards Kununurra, where David had his home. On the way, we turned off the main road again and dropped into a small mining camp.

Miners had been digging up the Kimberley for nearly a century. A gold strike near Halls Creek in the 1880s brought thousands of prospectors flocking from all over the country. It was a long and exhausting journey and most of those who made it were doomed to disappointment. Many of them later drifted down to the Pilbara. In more recent times, iron ore had been mined on Coolan and Cockatoo islands, just off the Kimberley coast. Then in the '60s there was a fresh outbreak of exploration.

The mining camp where we called was run by a large Canadian company, searching for copper. The head geologist was pleasant but uncommunicative.

'They won't tell you a thing,' said David as we drove away. 'Some of the locals reckon they have made a big discovery, but are not saying anything.'

That was to become a recurring theme in the Kimberley. Drilling teams came and went. The rumours came and went with them. During the 1970s the main subject of rumour was oil. The locals were convinced that the Kimberley was about to become the Texas of Australia.

Inevitably, the miners came into conflict with the Aborigines. It was the search for oil which eventually was the trigger, and when the confrontation came, the resulting storm was in the best Kimberley tradition.

Nookambah, the place where it happened, became the symbol of Aboriginal–white confrontation. It was a pity, because in the long term Nookambah was an example of Aboriginal achievement. A cattle station, Nookambah was bought by the government and handed over to the Aborigines to manage and develop. There were plenty of cynics who said it would collapse and be a humiliating defeat for Aboriginal aspirations.

I tended to think otherwise. Two of the Aboriginal leaders, Dicky Skinner and Friday Muller, had worked for us at the Fitzroy Crossing hospital and became health workers, who drove their own vehicles and looked after their own people. We had a great respect for them. Then along came an oil exploration company which made some discoveries that were significant enough to suggest that there were commercial possibilities. The company wanted to drill on the Nookambah station. The Aborigines said that the place where they wanted to drill was a sacred site.

The Western Australian government then intervened. Convinced that the Aborigines were being deliberately obstructive, it ordered the company to proceed with the drilling. By this time the media were well in control of the situation and everybody in Australia had a TV ringside view of the progress of the confrontation.

A long convoy of drilling equipment, heavily guarded by police, left Perth and headed north. Its progress was monitored each day by the media, as if it were a military column heading for the border of another country. Federal and state politicians were polarised in their attitudes, the federal government inclining in their sympathies towards the Aborigines.

As the column drew near Nookambah, Aboriginal elders and a few whites, including one of my colleagues, sat down on the track which led to the drilling site. When the police ordered them to move they refused. They were arrested and placed in the Fitzroy Crossing gaol.

By this time the weight of public opinion had swung heavily behind the Aborigines. The mining company was embarrassed and withdrew as soon as it could. The Western Australian government won the battle, but as they say, lost the war.

All this, however, was a decade after my first visit to the Kimberley in 1965. Then, the ominous clouds were just beginning to gather. The storm was yet to break.

A minor storm broke as David and I drove into Kununurra. Drenching rain washed the town which had not long been built. It had none of the barren brick suburbia which characterised the new mining towns of the Pilbara. Kununurra had been built among the trees and the roads wound around them like lazy rivers. We called in at the Kununurra hospital, which had been built by the AIM. The two nurses looked exhausted.

'Too many people come up here for the wrong reasons,' one of them said. 'They're either getting away from the city rat-race or an unhappy love affair.'

I found out later that one of the nurses fell into the latter category.

Unlike Fitzroy Crossing and Halls Creek, the cause of the nurses' exhaustion was not related to Aboriginal health. It was the effect of isolation and an oppressive climate on the new white arrivals. That afternoon the nurses had been called out to attend a young woman in her twenties who had committed suicide.

'It's not that uncommon,' a nurse told me. 'And there are many more who come close to it.'

The storm which hit Kununurra as we arrived had flooded the airstrip. The plane on which I was to depart stood patiently while the pilot walked the length of the strip, examining the surface carefully. In one hand he carried a large metal stake. Finally he stopped and looked back at the plane. Then he looked in the direction where he would take off and drove the stake into the ground.

As I sat in the plane and waited for take-off it occurred to me that the stake represented the moment of truth. If we were not off the ground by the time we reached it, there would be no tomorrow. The pilot raced the engines, released the brakes and we charged down the runway. I saw the stake disappear beneath us as we climbed into the sky.

The Kimberley region was fast approaching its moment of truth. It was a country in crisis and not a place for people with faint hearts or fragile tempers. Soon it would be no longer possible to ignore the plight of the Aboriginal people. The question was, had we left it too late? And the answer to that question was not entirely up to the white people of Australia. A lot depended on the Aborigines themselves and the extent to which they could control their own destiny.

One incident which happened early during my visit to the Kimberley told me more about Aboriginal character than anything else. It was Sunday morning in Fitzroy Crossing. The padre held a religious service when he came, but not when the nurses were busy, as they were this morning. Still, I am a creature of habit in many ways, so I asked the nurses if there was a church service held anywhere.

'Up at the mission,' they said. 'The Aboriginal people have one.'

So I borrowed a vehicle and drove up to the mission. The compound was deserted. I got out and kept walking until I came to a large bower shed. It was a simple structure, consisting of four posts which supported a roof made from the foliage of trees. The floor was bare earth. This was the Aborigines' church. The pews were planks of timber, nailed to stumps driven into the earth. There were about a hundred Aborigines present.

I took a seat at the back.

At the front of the church was a pine packing case which I presumed to be the altar. On either side of it, sat two Aborigines. One of them was plucking at the strings of a guitar. After a while I realised he was blind. The people were singing in a slow sensuous tempo. Later, this changed to the raucous chant of a well-known revivalist hymn. Then, one of the Aborigines stood up and began to pray in the language of his people. I think I understood him through the language of his feelings.

But even in a church service, I could not escape from the wretchedness of Aboriginal health. There was a great deal of coughing and spitting and I felt that everyone present must be suffering from tuberculosis.

On top of the packing case stood a blackened billy and an old saucepan. It struck me that they were to be used for holy communion. They were certainly different from the conventional silver chalices and plates. So were the contents. Instead of wine, the billy contained cold tea. In a saucepan were broken fragments of damper.

The service continued to be punctuated with coughing and spitting. The thought of drinking from the billy after a hundred tuberculosis patients was not appealing. I squared my Christian shoulders. Two of the Aboriginal elders picked up the billy and the saucepan and walked towards the congregation. They kept coming. The next moment they were standing beside me. I took the billy and drank the cold black tea.

Since then I have never felt the same about the insipid grape juice dealt up in most churches. Neither have I felt quite the same about the Aborigines. They had more to give me than a possible infection.

3

A Spot on the Map

'Australia is the lowest, flattest, driest, emptiest continent on the face of the earth.'

I looked down at the bare landscape over which we were flying. It was stretched out between the horizons in an unbroken brown monotony. Nothing I saw gave me evidence to disagree with the statement. But I couldn't let the comment go without question.

'Where did you pick up that shattering piece of information?' I asked the pilot.

She turned to me and grinned.

'In one of those magazines in a doctor's waiting room,' she answered. 'But after flying out here a few times, I didn't take much convincing.'

I looked down again, decided that I too would accept the evidence of my eyes and we lapsed back into that companionable silence which comes with long flights in little aeroplanes. Below us the landscape reeled off in slow motion. We had been flying for a few hours and there was still a long way to go. Four thousand feet below us, a narrow ribbon of dirt track pointed the way ahead. In the distance, a small smudge of dust moved slowly along the track. Someone was driving out into the desert to an unseen destination.

I shifted in my seat to ease a stiff muscle and said, 'I'd rather be flying over that vehicle than driving behind it, eating its dust for the next few hundred kilometres.'

It was mid-afternoon and the outback sun, tiring of its daily blaze of glory, was beginning to lower itself into the haze of the horizon. A long way ahead of us, its rays were turning the salt bed of Lake Frome into a tinsel strip of shining silver.

Nearer to us, on the other side of the plane, there was a small glitter of reflection. It was the iron roof of a homestead. As we came closer we could see a handful of buildings; the homestead itself, with a few trees close to it and two large sheds a short distance away. There was no sign of life.

I looked forward again. The smudge of dust we had seen was now travelling at right angles to the main track and disappearing towards the north. Shortly it would be gone. So would we. It was like looking at a vast stage, empty of everything except a few props in the form of the tiny buildings. There were no actors and shortly the audience would be gone.

'The family that lived down there has a few problems,' I said, pointing down to the homestead. 'The woman is in hospital in Adelaide. It's pretty serious.'

The pilot looked down at the lonely homestead. 'It's a hell of a long way to go hospital visiting at night after work.'

It was in fact 700 kilometres from the homestead to the hospital in Adelaide. I looked down again and wondered for the millionth time how people coped with this kind of isolation. The homestead is a spot on the map of Australia and it has to be a pretty large map if you want to find it.

There are many similar spots on the map of the outback. Each of them represents a tiny handful of people who live in extreme isolation. Some of the spots are well known, despite the fact that they have only a few people and no apparent importance. Names like Birdsville and Oodnadatta are household words in Australia.

In 1974, nine years after my first exposure to the outback, I became Superintendent of the Australian Inland Mission. Visits to well-known spots on the map, such as Birdsville and Oodnadatta, became part of my pattern of life. On this trip we were headed firstly for a place called Andamooka. Then we would fly north to Oodnadatta and from there across the Simpson Desert to Birdsville. At each of these centres we operated a nursing outpost hospital which belonged to the tradition of Flynn of the Inland.

But the days of camel trains and Flynn's old Dodge buckboard vehicle had long since gone. The fastest and most efficient way to travel around the outback is to fly. But the plane must be small and the pilot very experienced. By the grace of God and good friends, I had access to both. So whenever I had to visit a series of spots on the map, I began by picking up the phone.

You can always tell a bushie on the phone. The voice is cheerful and unhurried. It was a bushie who answered my call.

'Where do you want to go this time?'

'Andamooka, Oodnadatta and Birdsville. Can you spare me a week?'

We discussed times and dates and then the voice at the other end said, 'If you can fly to Broken Hill, I'll meet you there.'

My friend was probably already wondering how to organise things around the station to get away for a week's flying, but in due course I flew to Broken Hill. When the Fokker Friendship landed, I stepped out into a world of vaster space, bluer skies and cleaner air than I had left in Sydney. I stood for a moment and drank it in. My fellow passengers were quickly whisked away into town. Those who were flying back to Sydney watched the captain walk fussily around the plane, kick the tyres and inspect the propellers.

After a while, a tinny metallic voice came from the terminal building and invited the outbound passengers to board. They drifted across the tarmac like cows heading for the milking shed. The door slammed shut and the Fokker taxied out for take-off. Soon it was speeding down the runway, clawing its way into the sky and folding its spindly legs into the underbelly of the engines.

In a moment the sky was empty again and silence settled over the airport. I strolled across to the shade of a big hangar and waited.

The little plane arrived unheralded. There was no roar. It simply appeared at the far end of the runway and put down. I stood and watched it run along the strip and turn neatly into the parking area. The pilot taxied over to the fuelling bowsers and switched off the engine. The cockpit door opened and a woman stepped out.

Leonie lived on an outback station and loved flying. She was a disciplined pilot who had a healthy respect for the vagaries of the weather in the outback. As we greeted each other, a large figure emerged from the shadows of the cavernous hangar and came towards us.

Howie ran an outfit called Barrier Air Taxis. He sold fuel, repaired planes, gave advice and would fly you anywhere from Birdsville to Beltana. His face was lined with care and kindness. We chatted about the weather and the strengths and weaknesses of our little plane as he filled it with fuel. Then we jumped in and took off.

Two hours later we were flying over the lonely homestead where the woman had gone to hospital in Adelaide. Later still we were flying abeam of Lake Frome. At close range, the reflected light from its salt bed was so blinding we could not look at it. Instead we concentrated on a mass of dark ominous lumps looming up on the horizon. We were approaching the Flinders Ranges. It is not a high range, but you wouldn't want to be crossing it on a rough night in a small plane.

We crossed and made a sharp descent into Leigh Creek. Of all the places I have seen from the air, Leigh Creek is the ugliest. It is the site of an open-cut coalmining operation. The coal is railed south to supply South Australia's biggest electricity power station, while the mines look like deep black scars which have congealed on the blanched, barren face of the earth. The original town of Leigh Creek was almost as unattractive. It was built in the knowledge that one day it would be pulled down to mine the coal seam that lay underneath. We refuelled at Leigh Creek and took off for the final stage of the flight to Andamooka.

Ahead of us lay another large expanse of salt, Lake Torrens. Somewhere on the other side of it was Andamooka. We began to scan the horizon.

'It looks like the surface of the moon,' Leonie observed.

The earth below us was an orange colour. But there was one area which had a rash of small eruptions. They were white blobs with black centres and the earth was pockmarked with them. It looked as if a mysterious disease had broken out.

In a sense it had. Andamooka had been born of a disease and its name was opal mining.

Each black hole represented a mine shaft. The white circles were mounds of clay which had been thrown up out of the shafts by the miners. There were other larger cuts where miners who had become impatient with underground mining had used bulldozers to scrape out the earth.

I looked about for the landing strip.

'According to the map, there are two strips,' said Leonie, 'but I don't know which to use. One is close to the town and the other is about twenty Ks away.'

The answer to our dilemma came like a sepulchural voice from heaven. The plane's radio gave a preliminary crackle and then a voice with a heavy European accent said, 'Aircraft over Andamooka, land on the town strip'.

It was a bit like God giving Moses the Ten Commandments. We approached the strip nearest the town and landed. Later we learned that the voice belonged to a man called Rudi Duke. He was the self-appointed air traffic controller for Andamooka and his Good Samaritan role had saved many confused pilots from serious mishap.

We also learned later that the strip on which we landed was not 'the strip of choice' for discerning pilots. It was built on a rise, covered with rubble and subject to treacherous cross-winds. The other strip was larger, flatter and smoother, but when the rain made its occasional visit to Andamooka it turned the Clay Pan Strip into a dangerous quagmire.

At one end of the town strip was a solitary wind sock swinging listlessly from a post. Alongside it stood an old combi van. Alongside the van stood two women. The entire staff of the Andamooka hospital had turned out to meet us.

We touched down on the rough rubble strip and taxied towards the combi van. The two women came across as the pilot switched off the engine. We climbed out and exchanged introductions.

Mary, one of the nurses, gave the van a sharp kick. 'This vehicle,' she proclaimed, giving it another kick, 'is a bomb.'

We threw our bags and ourselves into the back. The two nurses climbed into the driving cabin. One of them took control of the wheel and the other took control of the conversation. She gave us a lecture on the faults and failings of the vehicle. The nurse who was driving gave a practical demonstration of the lecture.

The van slipped and slithered up and down the sharp rises in the road. It jarred and jolted over rocks which littered the track. On the slightest rise the van developed a violent attack of automotive asthma. Finally, on one rise, it gave up altogether.

We got out and pushed.

We bounced along a little further until one of the tyres suddenly collapsed under the strain. The two nurses looked at me accusingly, as if I were personally responsible. I got out and changed the wheel.

It should be explained that in 1974, almost all the outpost hospitals were without a vehicle. The nurses were dependent on the goodwill of the local community for transport. The combi van was Andamooka's 'on loan' contribution to the work of the hospital. The nurses were giving me a powerful argument for a vehicle of their own. By the end of the journey, I was virtually battered into submission.

The hospital was a white weatherboard, no frills building set on a bare hill overlooking the town. I looked down on the conglomeration of shacks which made up Andamooka.

A regular occurrence in the outback. Changing a tyre on the Andamooka ambulance.

No prizes for town planning. The Andamooka opal mining community in outback South Australia.

Just a winch and a rope. An opal miner at Andamooka prepares to go underground.

Birth place of the AIM. Oodnadatta Hospital built by Flynn of the Inland and opened in 1911.

They were constructed of corrugated iron and any other material which happened to be at hand. There were no made streets. Rough tracks wound down from the rises on which most of the houses were built. The tracks appeared to have been built along the lines of least resistance. There were one or two larger buildings at the bottom level, which I gathered was the town centre.

On the rising ground to one side of the hospital were two large concrete tanks. I nodded towards them and said, 'How's the water supply?'

The question, asked in innocence, provoked another flow of rhetoric from the nurses, both of whom were of Irish descent.

'One of my ancestors was deported from the old country for stealing bread,' said Carmel. 'Someone has stolen our water and if we can ever prove who did it, I'll have them deported back there.'

Andamooka and the surrounding region has the lowest rainfall in Australia. There are times when even if the residents collected every drop of rain that fell, it was not enough. Then water has to be brought in by tanker. At the time of this visit, Andamooka had been passing through a particularly serious drought.

'We had been very careful,' said Carmel. 'One of our tanks was still practically full. Then one night we decided to go to the outdoor pictures. We left the usual, 'Sisters at Pictures' notice pinned to the front door of the hospital and headed off. After the show we invited a few people back for a cup of tea.'

'I turned on the tap,' said Mary, taking up the story. 'It piddled for a second and then nothing. We went outside with a torch, climbed up the ladder and shone the torch down inside the tank. It was bone dry. Then we looked round the base to see if the water had leaked out. The ground was as hard as a rock.'

While the nurses had been enjoying the pictures, someone had stolen 4000 gallons of water. The nurses went and reported the matter to the police.

'Well,' said one of the young policemen, with a grin, 'can you describe the stolen property?'

The nurse nearly exploded. 'How do you describe 4000 gallons of bloody water? It's not like a car or a TV set.'

Outback places have their own networks. Within a day, everyone knew who had stolen the water, but there was no way it could be proved.

The social gathering point for the people of Andamooka was a place called the Tuckerbox. The nurses took us there for dinner. From the outside the Tuckerbox was no more impressive than any other building in the town. I had the feeling there should have been a hitching rail outside and the impression of a Wild West saloon was confirmed when we went inside, except the customers were miners and not cowboys.

There were tables for drinkers, tables for eaters and tables for billiard players. We sat down at one of the tables for eaters. Some of the other customers greeted the nurses with cheerful friendliness, but the looks they cast in my direction were less than welcoming.

'The boss from Sydney,' said one of the nurses, by way of introduction. Then she turned to me and said in a more confidential tone, 'They probably think you are from the Taxation Department'.

The nurses pointed out some of the characters of the town; German George, Belgian

Freddie and others. The miners were of all ages and nationalities. There were a few family groups and I gathered that eating at the Tuckerbox was the Andamooka equivalent of MacDonalds or the Pizza Hut, for the kids.

After a while, some of the miners came and sat with us. They asked what I was doing up here. I explained that I had to keep an eye on the nurses. That led immediately to a raucous exchange between the nurses and the miners with allegations and denials of wild behaviour.

One of the miners asked if I would like a game of snooker and we played a few games. He was very polite about my skill and bad luck, but I suspected he was keeping something in reserve. He won easily. I went back to the table and he wandered off in the other direction.

'He's a nice bloke,' I commented. 'Who is he?'

'We think he's a murderer, on the run from New Zealand,' said one of the miners, 'but that's his affair.'

Living in one of the spots on the map of the Australian outback is a bit like living in a crucible of humanity. Every step you take, every breath you inhale, is observed by someone and placed under the microscope of human curiosity. But the past is another country. You didn't pry into people's past. The present and its difficulties are enough in the outback.

Some of the miners told me about life in Andamooka. Like most strikes of its kind, the mining of opal at Andamooka began in a blaze of publicity. As some of the miners were not backwards in celebrating their successes, it was a case of 'drinks all round' and sore heads for several days. Then there was a spate of robberies and the theft of some big opal collections. The celebrations suddenly dried up and there was no more talk of big finds.

'Anyway,' said one of the miners, 'the day of the big finds has finished. A few of these fellas,' looking around the room, 'go for months without finding anything and I reckon have just about exhausted their capital.'

'Why do they stay?' I asked.

'Why do people buy tickets in lotteries?' He looked at me as if I were stupid. 'In any case, it's a way of life with which they are comfortable. It doesn't take much to live and the thought of going back to the city frightens the life out of them.'

Across the room, two men sat at a table by themselves. Nobody went near them. They said little to each other. Occasionally, one of them would go and buy more drinks. They looked like some husbands and wives I have seen, sitting together in the comfortable bonds of boredom. I nodded towards them and asked one of my new friends who they were.

'The fellow on the left is Kurt. He's a miner. The other bloke is an opal buyer from Hong Kong. We reckon Kurt's got some opal to sell. He won't give it away, but the buyer has his price too. He's been up from Adelaide for a couple of days. I reckon he'll be gone by tomorrow.'

'Will Kurt sell?'

The miner put down his glass and looked across at the two men who were as animated as a couple of chess players.

'We'll never know,' he answered.

The miner with whom I was speaking was a young man called Stephan. He told me he had been a surveyor in Yugoslavia, but had caught the opal-mining disease when he came on a visit to Andamooka. He invited me to visit his mine the next day.

We drove out to Stephan's mine the next morning. The track wound round mounds of white clay and we passed several shafts with a solitary miner preparing to go underground. Eventually we pulled up alongside a digging, got out and walked across. There was a windlass placed over the shaft. I looked down. The shaft was very narrow, hardly wide enough for one man to wriggle down. On one side of the shaft, a narrow ladder was suspended. The bottom of the shaft appeared to be about fifteen metres below the surface. There was no-one in sight.

The nurse who came with me called down the shaft. 'Are you there Stephan?'

There was a muffled response and eventually a head appeared from a tunnel at the bottom of the shaft. Stephan wriggled sideways out of the tunnel and began to climb the narrow ladder. He brushed the dirt from his clothes and asked if I would like to go down.

'Is there room for two down there?' I asked. 'Or do I fly solo?'

Stephan assured me that there was room for both of us, although it would be a bit 'cosy'. I climbed down the ladder and found that there was a narrow tunnel driven off to one side. Somehow we both managed to crawl in. It was indeed cosy!

'We find the opal at about this level,' explained Stephan. 'It was a sea bed originally and the opal is in crustaceans that lay on the bottom. We put the shaft down this far and then just pick out a drive like this one.'

He wriggled onto his side. 'You have to pick carefully so as not to chip any opal.'

As he was speaking, Stephan wielded a small pick with considerable dexterity. I asked if he had found any opal lately.

'I haven't had a strike for a while. Perhaps you will bring me luck.'

The pick struck something hard and he began to scrape away the clay with his hands. Then there was a sharp intake of breath.

'You can come again,' said Stephan.

It was not a big strike, but there was enough opal to bring a smile to his face and to mine, too, I must confess.

We climbed up the narrow ladder to the surface and triumphantly showed the nurse what we had found. She tartly remarked that if I could do that for a miner, I might be able to discover a new vehicle for the nurses.

That night at the Tuckerbox, Stephan was offering to sell my services as an opal diviner. We were sitting at a table and yarning with some of the miners. I asked them about the dangers of mining in such confined spaces and with most of the miners working by themselves.

'What would happen if someone had a heart attack?'

'Someone did,' one of the miners replied calmly.

'Yeah, and that's when Annie broke her vow,' said another.

Annie had been a nurse at the Andamooka hospital and her story explains why the nurses were held in such high respect. Shortly after she arrived in Andamooka, one of the miners took her out to the diggings and asked if she would like to go underground.

She looked down the narrow shaft and shuddered. 'I hate confined spaces,' she had said, with some heat. 'You wouldn't get me down one of those shafts for all the opal in the world.'

A month or so later, she was working in the clinic when a car drove up to the hospital front door with a screech of brakes and the slamming of doors. A miner burst into the clinic.

'It's Harry!' he panted. 'He's lying at the bottom of his mine and I think he's dying.'

Annie shouted to the other nurse that she was going out to an emergency and to stick by the radio so she could let her know what was happening. She and the miner drove at breakneck speed along the rutted track that led to the fields. When they arrived at Harry's diggings Annie clambered over the pile of dirt without hesitation, stepped onto a rung of the ladder and began to clamber down the shaft. She reached the bottom and bent over Harry, who was lying unconscious. He had had a heart attack.

Working in the confined space at the bottom of the shaft, Annie managed to revive him. She yelled to the miner at the top to radio the nurse back at the hospital and tell her to call the flying doctor for an emergency evacuation. By this time, other miners had arrived on the scene. With a lot of difficulty, they managed to bring Harry up to the surface. He lay on the ground in the strong sunshine, alive but obviously dangerously ill. The nurse from the hospital arrived to say that the flying doctor plane was on the way.

They put Harry on the back of a truck and proceeded to drive carefully to the airstrip. One of the nurses rode on the back with Harry to keep an eye on him. Then about half way to the strip, she pounded on the driver's cabin.

'Stop the truck!' she yelled. 'He's having another attack!'

The truck stopped abruptly and they lifted Harry onto the ground. Once again the two nurses administered heart resuscitation and, after some strenuous effort, brought him round. They reached the airstrip and waited anxiously until the plane arrived. It came equipped with resuscitation equipment and flew Harry to Port Augusta and the base hospital.

'Did he survive?' I asked.

'Oh yes,' replied the miner casually. 'He's back again, you know, working in his mine.'

'They're not all that lucky,' chipped in one of the other miners, putting a fresh glass of beer in front of me. 'You remember Faye?'

Yes, I remembered Faye. She had been one of our nurses at Andamooka when the hospital had been first opened. Her husband had been killed underground, when his mine had collapsed, and no-one had known about it until too late.

'It's why a lot of them are moving away from underground work. They're bringing in bulldozers and scraping the earth down to the level where the opal is.'

A few tables away from us, a family was having dinner. There were two children, a girl and a younger boy. One of the nurses noticed me looking at them.

'That's Wayne,' she said, nodding at the little boy. 'He caused us some trouble too.'

She told me that Wayne and Julie had been left in Andamooka while the parents went off to Adelaide for a few days. Although the girl was not as old as all that, she was pretty sensible. The night after the parents left, Julie brought her young brother up to the hospital.

'He says he's got a bad tummy ache,' she reported to the nurses.

They examined Wayne, but could find nothing wrong. Suspecting it might be a case of missing mum, they admitted him overnight and made a bit of a fuss over him. Next morning he was still complaining of pains in the stomach. The nurses examined him very carefully and contacted the flying doctor at Port Augusta. He listened to their report and commented that he could not think of anything they had not done. The nurses made contact again in the afternoon and reported that there was no improvement.

'I don't think there's any cause for alarm,' said the doctor, 'but I can see you're worried. So we'll send the plane and evacuate him.'

The nurses waited until the radio crackled with the message that the plane was well on the way and then drove the boy to the airstrip. It is a lonely place to wait when you have a sick child to look after. The plane landed and took off again with the boy.

Several hours later, the doctor made contact.

'It was just as well,' he said. 'It was peritonitis. Another hour or so and we would have lost him.'

I looked across at the family happily eating their huge Tuckerbox sandwiches. When they had finished and were leaving, they walked past our table and the parents gave the nurses a warm smile. I understood why. I also thought that it's not every place in Australia where the hospital staff see their ex-patients on a regular social basis.

We left Andamooka the following morning. In the time we had been there, I had not been allowed to forget the iniquities of the infamous combi van. But as we drove out to the strip I became aware that the vehicle was not bucketing to the same extent as on the trip into the hospital when we had arrived. I looked out the window and saw to my astonishment that the road on which we were travelling was singularly lacking in sharp dips and rises and had been recently graded.

'Where did this miracle of road-making spring from?' I asked suspiciously.

'Oh, we had a visit from the state premier a few weeks ago,' replied Mary, airily. 'And they built this road especially for him.'

'But don't forget you promised us a new vehicle,' chipped in Carmel.

We got into the plane, rumbled along the rubble airstrip and wavered into the air. I looked down at the combi van and the two nurses waving us goodbye. The clay mounds of the opal fields and the scattered shanties of the town passed under our wings. Ahead was the broad sheet of shining salt that was Lake Torrens. In every other direction was desert.

I thought that if the government could lay down a new road for the premier's 'once only' visit to this out-of-the-way place, then people like Harry the miner and Wayne, the little boy with the peritonitis, deserved to be evacuated in something better than a beaten-up combi van.

4

The End of the Line

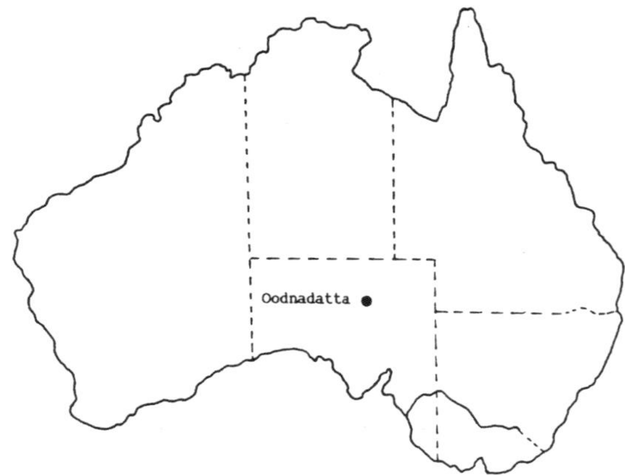

'There's the lowest point of the lowest continent in the world.'

Leonie was continuing her doctor's waiting room geography lesson. I looked ahead and saw a broad sheet of salt lake beginning to appear on the horizon.

'It's Lake Eyre and it's twelve metres below sea level.'

'I've flown over Lake Eyre before,' I replied, not wanting to appear totally ignorant. 'It was on a commercial flight from Adelaide to Alice Springs. And it looked a little different from that. It looked like an inland sea.'

For decades on end, Lake Eyre was a flat dry bed of salt whose only use was to help futuristic racing cars break land speed records. But in 1974, heavy rains in the north poured through the Channel Country of Queensland and emptied into Lake Eyre, filling it to overflowing. It had been a spectacular sight. But it also became a severe frustration to people travelling north, especially by train. The line was washed out for weeks.

We had left Andamooka and were heading north, straight up the centre of Australia, to Oodnadatta. From the air, the earth around Andamooka had been a rich red. Further north it turned into a much lighter colour, as if the constant harassment of the sun had caused it to fade, like a badly worn carpet.

Our flight path crossed a number of dry creek and river beds running from west to east. They came out of the desert country to the west and stretched away to the horizon, eventually emptying into Lake Eyre. They looked as if the long journey had dried them out in exhaustion.

About halfway up the track to Oodnadatta, Leonie pointed down and said, 'That will be William Creek.'

There was a line to the right of our path, which ran as straight as a ruler from the southern to the northern horizon. It was the railway line from Port Augusta to Alice Springs. A solitary iron-roofed building close to the line, marked the site of the William Creek pub.

Nearby, was an area which was even more bleached and barren than the surrounding country.

'That'll be the race track,' Leonie observed.

The annual William Creek Races drew hundreds of people from all over the countryside. Looking down on the barren empty country with no sign of life, it was hard to understand why.

Two hours after leaving Andamooka we were preparing to put down at Oodnadatta. In 1911, when Flynn of the Inland built his first hospital, Alice Springs had barely begun and Oodnadatta was the most important town between Port Augusta and Darwin. The railway line north from Port Augusta only reached as far as Oodnadatta, which was the end of the line in more senses than one.

Journeys further north were made by camel team or horses. The Oodnadatta hospital was the only medical facility over a span of nearly 4000 kilometres. Looking down on the country, I began to appreciate the importance of what Flynn had done.

From the air, Oodnadatta looked as if it hadn't changed much in sixty years. But the airstrip was good and that was an improvement. It would have taken Flynn a week to make the trip that had taken us two hours.

We landed and taxied towards the wooden building that served as a terminal. There was nobody in sight. Then I saw a car parked under the shade of a tree. As the pilot switched off the engine, the doors of the car opened and four people got out. Two of them I guessed were the nurses. The other two were policemen. I had a horrible thought that the South Australian police must have invented aerial speed traps. As it turned out, I was not far wrong.

We introduced ourselves and I made some fatuous remark about the lovely weather they had turned on to welcome us. One of the policemen replied in a tone something less than sunshine.

'This vehicle of the nurses,' he said, pointing to the old Holden from which they had emerged.

'What about it?' I replied, sensing trouble.

'It's unroadworthy. If you don't do something about it we'll have to ban the nurses from driving it.'

I looked at the nurses reproachfully. They were looking at the sky as if they were above such mundane disputes, but I thought I detected a look of amusement and triumph flit across their faces. I promised to look into the matter. We squeezed into the car and drove back to town.

The main street of Oodnadatta is a wide and empty strip. The railway line runs along one side to keep it company. The railway station is an old, elegant, red-brick building, which reflects the glories of a bygone era.

The buildings on the other side of the street were not so glorious. They were spaced at a distance from each other. None of them looked as if it would qualify for National Trust classification, but then you never know these days. Opposite the station was a squat building which, according to a sign, claimed to be the Transcontinental Hotel. Outside it, a couple of dusty vehicles were angle parked. I suspected that their owners were angle parked at the bar inside.

Further along was another unprepossessing building. It was the general store. Several Aborigines stood outside, looking as if they regretted giving it their patronage.

The last building in the main street, before it emptied you back into the desert, was the Oodnadatta hospital, built by Flynn of the Inland. I had seen many photos of it. The real thing was unmistakable, with its high-pitched, corrugated-iron roof and the big old palm trees planted outside.

We pulled up and got out. I would not have been surprised to see Flynn himself emerge from the front door. I looked back down the long wide empty road. The midday sun beat down mercilessly. There was not a soul in sight. Oodnadatta had closed for lunch. We went into a large friendly kitchen with a big table in the middle. Without being asked, everyone, including the two policemen, sat down at the table. One of the nurses put the kettle on the stove. The Oodnadatta hospital was open for business.

Since that day, I have sat at the table in the kitchen of an outback hospital a thousand or more times. I came to learn that the kitchen table at an outback hospital is a bit like the communion table or altar in a church. Sooner or later, everyone comes there, looking for refreshment of one kind or another. For the moment it was two travellers and two policemen, happy to accept a cup of tea.

The nurse passed round the cups and sat down. Somewhere at the other end of the hospital a bell rang sharply. The nurse got up again and disappeared down a long passage.

'What are the clinic hours?' I asked, guessing that it was the clinic door bell that had rung.

'There aren't any,' replied the other nurse. 'Oodnadatta isn't a nine-to-five kind of town. Station people travel long distances to get here. You never know when they will arrive.'

The nurse got up to refill the teapot.

'Then there are the men who work out in the desert in the drilling camps. They disappear for weeks at a time and only come in when they have finished their job.'

'What about the Aborigines?' I enquired, knowing they constituted the great majority of the patients.

'They don't have clocks,' said one of the policemen.

The nurse who had gone to answer the clinic door returned and sat down. The bell rang again. She got up and went back to the clinic.

By this time, the two policemen were treating me as something less than a criminal. They told me that they worked closely with the nurses because they were often involved in the same situations, such as fights and road accidents. I suspected that they were also not averse to some pleasant female company.

We talked about the vehicle and its problems. It had certainly seen better days. General Motors kindly gave the AIM every millionth Holden that came off the production line and this one was the two millionth. It had already done rugged service in Central Australia before coming to do 'town work' in Oodnadatta.

The kitchen door opened and a woman came in and sat down at the table. She was obviously a regular visitor.

'Where's Ben?' asked one of the policemen.

'He's up north of Macumba,' replied the woman, whose name was Robin. 'There's an exploration team working out in the desert up that way. I don't know how long he'll be away.'

Ben was in charge of road maintenance for the region. He had some of the roughest country in Australia to contend with and was generally away for weeks at a time. Robin lived at the other end of the main street. She had come down to see if one of the nurses would be at the Progress Association meeting that night.

'We're making final arrangements for the gymkhana,' she explained.

One of the nurses promised to attend.

'We get involved in all sorts of things here,' she explained after Robin had left. 'Julie is the secretary of the Parents and Friends at the school. I go to the Progress Association meetings.'

The door to the kitchen from the outside opened again. The man who came in looked as if he had been out in the bush for six months. I glanced at his vehicle through the open door and amended my estimate to twelve months.

'Got a job for you, sister,' he said cheerfully, stretching out his arm. An ugly and angry red weal ran almost the length of his forearm. It was obviously infected.

'My God!' said the nurse, getting up quickly. 'How long has it been like that?'

'About a week or so.'

'Why didn't you come in before?'

'We weren't due to come in until today.'

The other nurse let out a sigh as the injured man was led off to the clinic.

'They never learn,' she said. 'They think they have to tough it out.'

The afternoon wore on and the day died as it does in the desert, suddenly and without warning. I felt the cold hands of the night making mockery of the light clothing I was wearing. The temperature at Oodnadatta has been known to drop to 8 below zero in the winter.

'I hope you have a nice warm bed for me,' I said.

'We've put you in the maternity ward,' replied the nurse.

Everyone laughed.

We went to bed early and I was soon asleep. I woke to the sounds of Armageddon or some similar apocalyptic event. There was a great racket going on and a lot of people talking at the top of their voices. Most of them were Aborigines. Then there was a loud banging on my door and a voice demanding I should get up immediately.

'You've got to get out!' yelled one of the nurses. 'We have a woman in labour.'

I looked at my watch. It was 2 a.m. Children set out to defy their elders even at birth. I struggled out of bed into the bitter cold of an Oodnadatta night.

The nurses went about their work with calmness and efficiency. It was a rule that nurses who served in hospitals like Oodnadatta had to be qualified midwives. At that time, Aboriginal women in many outback centres didn't come near the hospital until they presented in labour. The Flying Doctor at Port Augusta was as far away from Oodnadatta as Melbourne is from Sydney. The possibility of him getting to Oodnadatta at 2 a.m. to attend an Aboriginal woman in labour was minimal. It was the nurses or nothing.

Soon there was an additional noise to the existing orchestration of the voices of the Aborigines who had come with the expectant mother. A baby had been born; an Aboriginal baby. I caught myself wondering what the future would hold for this child. When it was all over, the nurses found me another bed.

Next morning they confessed that they had been warned on the bush telegraph that the mother might come in last night.

'We felt it would be good for you to have a first-hand experience of how we work,' they said, trying to keep the smiles from their faces.

As we sat eating breakfast, the kitchen door opened and a small woman in a yellow uniform dress came in and sat down. She had a quiet composure about her that spoke of mature efficiency.

'You know Adele,' said one of the nurses.

Yes, I knew Adele, at least by reputation. She had served in a number of outback hospitals, working without fuss but with great effect. Her present role was to improve the health of the Aboriginal people in and around Oodnadatta. It was a tough challenge. There were camps of Aborigines on a number of the stations in the region and Adele visited them regularly.

After breakfast, I went with Adele as she took her morning walk around town. It seemed like more of a social visit than a tour of medical inspection. As we stopped and chatted with Aborigines in the street, I couldn't help comparing this with the work of a nurse in a large city hospital, bustling around the wards, taking temperatures, making notes on charts and adjusting drips. By contrast, Adele's work was bush track medicine.

We crossed the railway line and went towards a group of cottages where some of the Aboriginal people lived. Outside one of them, an old lady sat in the sunshine. There was a bundle of bedclothes on one side of her and a small camp fire on the other. She was wearing a bright cotton print dress and a knitted beenie on her head. Her smile was wide and open. Adele squatted down beside her and they chatted for a while.

'How are those eyes?' asked Adele.

'Good!'

Adele bent over and examined them. A couple of dogs bounded across and joined the party. The sun began to gather strength in the sky. Adele straightened up, apparently satisfied with what she had seen. After a while we said goodbye and walked away.

'She had trachoma,' Adele told me as we trudged the dusty track. 'The Trachoma Team came out and fixed her. I was left to continue the treatment. The dust and the wind and the sun don't help. I'm not sure how long the treatment will last. Eye diseases run riot out here.'

We passed a couple of young Aboriginal girls returning from the town.

'Hello, Sister,' they giggled.

'I've got problems there,' said Adele. 'They get pregnant very young, sometimes as young as twelve or fourteen. Getting them to be responsible about contraception is not easy.'

We continued our walk around town. An Aboriginal man stopped to talk to Adele. He told her he was not feeling well.

'Better get up to the hospital today,' she suggested. 'Hospital sisters will fix you up.'

As we walked on she remarked that by encouraging people to go to the hospital early, many of the illnesses were treated before they became critical. Then we turned into the yard of a small house and walked up to an old Aboriginal man who was sitting outside. His face was lined with laughter and his eyes shone with happiness.

'Have you taken your tablets this morning?' Adele asked with mock severity.

The old man grinned like a little boy. 'I forget, sister,' he answered, with a look of comic repentence.

'You're teasing me,' Adele said.

'He's diabetic,' she told me later, by way of explanation. 'And we have to watch his medication carefully. But he's a wonderfully wise old man and he has taught me a lot about the people. So I don't make as many mistakes.'

Later in the morning, I noticed that the nurses back at the hospital were getting ready to go out. 'It's the day for the Ghan,' one of them told me as they walked out the door.

I walked out with them and we strolled down the main street towards the station. There were other people moving in the same direction in an unhurried way. We crossed the line and went onto the platform. The nurses chatted to the other locals who had gathered.

The Ghan was the weekly train which came from Port Augusta in the south and passed through Oodnadatta on its way to Alice Springs. I went to the edge of the platform and looked down the line. In the distance I could see the train approaching. Moving slowly and seeming very tired, the Ghan consisted of a number of carriages for passengers, some goods trucks and a line of flat tops on which cars had been parked. In 1974, the road to Alice Springs was still unsealed and often in bad repair. Many people preferred to freight their vehicles on the Ghan, rather than risk the road.

The train moved slowly into the platform and the engines hissed to a halt. Behind them, the carriages and trucks jangled and jolted in protest. There was no rush and flurry of activity. A few passengers got out to stretch their legs. They looked across the line at the straggle of buildings that was Oodnadatta and decided that there was nothing of interest to see. So they turned their attention to the railway staff who were performing their rituals with spanners and water hoses.

The group of people from the town continued to chat to each other as they waited for their packages and parcels to be offloaded. Nobody was in a hurry. Time stood still. A small Aboriginal boy wandered down the street across the line. He stopped for a moment and stared at the train and the people who were staring back at him. Then he scampered off.

Gradually, the minor air of expectation the arrival of the Ghan had generated slowly dissipated. The vast emptiness of the land and the sky once again assumed control. As if in recognition, the engine let out a long sigh. The passengers disappeared back into the train. The station master waved a languid flag. Then with a weary grunt and the clanking of its aching joints, the Ghan resumed its journey north.

'We've got a few things to discuss with you,' said one of the nurses, later in the day, 'so we thought we'd have a talk after tea tonight.'

People continued to come to the hospital during the day, sometimes announcing their arrival with the strident bell at the clinic door and sometimes simply walking into the kitchen. One of the nurses was on duty and the other was kept busy doing the housework and cooking. Occasionally the nurse on duty would come out of the clinic and have a consultation with her partner. With the nearest doctor 500 kilometres away, the importance of their support for each other was obvious.

We sat down to tea and were joined by Adele. I asked if they ever got a break.

'One of us has to be here all the time,' was the reply. 'But occasionally, if the town is quiet, we go for a short drive out into the bush and just boil a billy. It's good to get out of town, even for a couple of hours.'

'That's how Adele learned to drive backwards,' said one of the hospital nurses, grinning at her. 'We went about twenty Ks out into the bush to collect some firewood and just have a stroll. It was getting a bit dark so we decided to go back.'

'I put the ute into forward gear and nothing happened,' Adele said. 'After a few attempts I discovered that it would only go in reverse. Fortunately we had a torch with us, so Chris sat in the back and shone it on the track while I drove in reverse for the twenty Ks.'

'And we only hit three trees,' Chris added.

We were just beginning to relax when the clinic door bell called us back to attention. The nurse on duty got up and went to answer it. We tried to maintain the frivolity of the incident with the vehicle and joked about other occasions when we had struck trouble driving. But raised voices from the clinic were difficult to ignore. One was the wail of an Aborigine in pain and the other was the nurse endeavouring to calm the patient sufficiently to administer treatment. After a while the nurse came back into the kitchen.

'Her husband hit her,' she said. 'They were both drunk.'

I asked if the drink was a big problem.

'It's the thing that gets to us most,' said one of the nurses. 'The constant stitching and patching up after fighting. It's largely the result of drinking. After you've stitched up a person for the third time in a week, your enthusiasm tends to wane.'

The clinic bell rang again. A nurse put down her knife and fork and got up. The conversation around the table began to wane. The nurse returned after a while and we tried again.

After several more disturbances we gave up trying to conduct a dinner party. I offered to lend a hand and clean up the mess in the clinic. At about 11 p.m. the woman who had come in earlier returned. Her husband had hit her again and there was another wound on her head. She was making an awful din and whenever the nurse approached her to treat the wound, she twisted away and howled even louder.

The nurse was quite tiny and normally a model of restraint, but it all became too much for her.

'Will you please shut up and lie still,' she shouted. 'If you don't, I might hit you myself.'

It had the desired effect.

About midnight we cautiously came to the conclusion that the night's drama might have ended. Thoughts of having a discussion had long since disappeared. But as we sat down for a final cup of tea, I felt compelled to ask if it had been worth their while coming to Oodnadatta.

'Yes it has,' replied one of the nurses. 'We get occasional nights like this, when you wonder what on earth you are achieving. But I love the people and I hate what the drink is doing to them.'

She looked across at Adele, who had stayed to lend a hand.

'Adele has got some good programmes going to prevent diseases and improve health. But it's going to take a long time, and in the end the people will have to do it themselves.'

Ordinarily that statement would have provoked a lot of discussion. But we were all too tired and the night was cold.

In 1914 the first nurse to work in the Oodnadatta hospital, Sister Bett, took a critically ill patient on a three-wheel rail trolley to the doctor in Port Augusta. The first 160 kilometres took eight hours and was made in the blazing heat, with temperatures soaring around 120°F (49°C).

Sixty years later, I travelled with Adele on a journey which took us over much the same route. This time we went by road and the vehicle was a utility. The main road south from Oodnadatta was still unmade and we had to make several detours to avoid deep washouts carved across the track. Eventually, we turned off the main road south and headed down a side track to a cattle station homestead.

We pulled up in front of one of the station buildings, which had most of its sides covered in flywire screens. The flywire door creaked open as we went inside and then banged shut behind us. This was the mess hut for the station workers. The floors and the tables were both made of bare boards. Station hands and stockmen sat around eating lunch.

At one table sat a squat solid man with a round florid face and eyes which had been permanently narrowed by the sun. He was still wearing his bushman's hat and I gained the impression that it didn't come off very often. The brim of his hat cast a shadow over his eyes and you couldn't tell whether they were mischievous or malevolent.

Adele was especially concerned with Aboriginal health, but she also looked after the white people on the stations when she was visiting. She walked up to the man wearing the hat.

'Gooday, Dick. This is the boss from Sydney,' she said by way of introducing me.

Dick looked me over without much enthusiasm.

'I suppose you want a feed,' he said, as if he wouldn't mind if we refused.

Later I was to learn of Dick's extraordinary generosity, but for the moment it had been a long time since breakfast. We said that a feed would not go amiss.

'It's over there,' said Dick, pointing to a table bearing large chunks of cold meat and a few salad vegetables.

We went over and helped ourselves, and then returned to Dick's table. There did not appear to be any knives and forks.

'Where's your knife?' Dick said aggressively.

I sat silent and looked helpless.

'You won't last long in the bloody bush without a knife. Go and get one over there,' he roared.

Dick Nunn had been the manager of Anna Creek station, which was one of the biggest pastoral stations in the country, for a long time and was well known and respected. He was a pretty down to earth character and was renowned for his hard, biting humour. I found out later that he was also the driving force behind the major efforts to raise money for the Oodnadatta Hospital and the Royal Flying Doctor Service.

On the north–south highway close to the Anna Creek station was the place called William Creek over which we had flown. It boasted a pub and a racetrack. Flemington need have no fear of a rival in William Creek. But once a year it was the venue for the annual William Creek Races, an occasion that almost rivalled the Birdsville Races in popularity and attracted people from all over the country.

The Oodnadatta Hospital and the Royal Flying Doctor Service have both benefited greatly from the proceeds of the William Creek Races and from a mammoth auction which Dick Nunn used to conduct during the course of the race meetings. Dick would auction everything in sight. On one occasion he was auctioning a lady's hat and his wife was bidding furiously for it before she let out a bellow of rage.

'That's my hat you're auctioning!' she shouted.

'Well, buy the damned thing,' Dick responded.

As we sat and yarned with Dick, I asked him if he had ever met John Flynn.

'Oh, yeah,' he replied, while continuing to eat his meal. 'He used to come and give us a little bit of religion and a little bit of medicine. He was an all right man.'

As we drove back to Oodnadatta later in the day, Adele summed up what I was feeling about Dick.

'He's a rough old blighter and I wouldn't like to get on his wrong side,' she said. 'But he'd do anything for the flying doctor and the hospital.'

The next morning we left Oodnadatta as we had entered, with a police escort. The young policemen who had been so amenable around the kitchen table resumed their official role. They repeated the long list of things wrong with the hospital vehicle. They said, with a meaningful look, that they hoped the nurses would have some good news soon. We shook hands all round before I got into the plane and left.

The Simpson Desert lay beneath us. We were on the way to Birdsville. It was another 500 kilometres of featureless country. With time to think, I began to contemplate the coincidence of being pressured for new vehicles by two teams of nurses in two hospitals, 400 kilometres apart. Or was it coincidence? The nagging nub of suspicion began to grow in my mind. Ahead of me lay Birdsville hospital, which had no vehicle at all, not even a battered old combi van or a museum-piece Holden. I had a sudden apprehension of what lay ahead and decided that the best method of defence was attack.

The rippled rows of red sandhills rolled beneath our wings as the Simpson Desert unfolded its majestic pattern. We passed over a whole spatter of salt pans, glistening in the sun. On the horizon was a long, dark smudge that indicated the course of the Diamantina River. Soon we were putting down on the Birdsville airstrip. We landed and taxied to the far end, where two women detached themselves from the boundary fence and walked across towards us. I jumped out of the plane and confronted them.

'Before you ask me for a vehilce, the answer is "No!" '

Their mouths dropped open in astonishment and their faces fell in dejection.

Then I grinned. 'But before we talk about it, come clean and let me in on the plot.'

'It was Mary at Andamooka who thought up the idea,' said one of the nurses, Beverley. 'We found that we could talk to Andamooka on the radio when the reception is clear. We can get Oodnadatta too. We all found that we needed vehicles badly, so Mary said she would soften you up by giving you a horror trip in the combi. Then the girls at Oodnadatta hit on the idea of bringing in the police. They'll do anything for the nurses over there.'

'As soon as you had left Oodnadatta,' the other nurse continued, 'they made contact and told us to close in for the kill.'

We all burst out with laughter. I had heard a lot about the bush telegraph. Now I had experienced it in action. The nurses wanted wheels and they wanted them badly. Their reasons must be good. Just how good, I came to learn.

5

We've Got Wheels

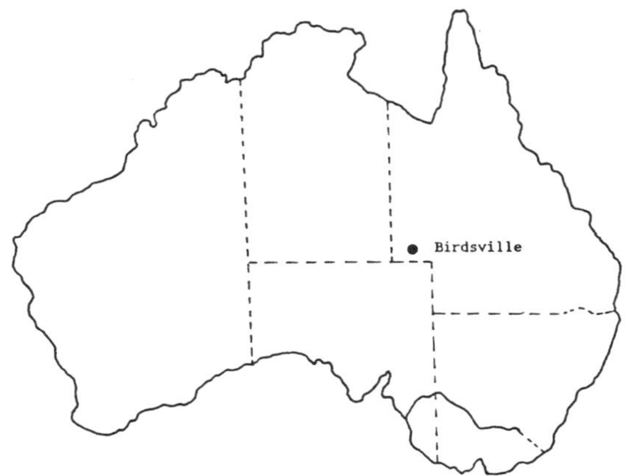

The nurses at Birdsville Hospital are used to improvising, and visitors there are expected to do the same. So after our confrontation about acquiring a vehicle, one of the nurses had the last word.

'Well, since we haven't got a vehicle, you'll have to pick up your bags and walk.'

I picked up my bags and walked towards the boundary fence of the airstrip. Across the road in front of me was a building I recognised. It was old and rambling, made from stone, with an iron verandah running right along the front. The nurse walking with me caught the expression on my face.

'Yes, that's the Birdsville pub,' she said. 'And no, you haven't earned one yet.'

I protested that my interest was purely historical.

The main street of Birdsville was very wide and stretched from one horizon to the other. At one horizon, in the far distance, I could see a small cluster of buildings nestling in a clump of trees.

'That's the police station,' explained one of the nurses. 'When they throw someone out of town, it's a long walk back.'

A long road to anywhere. Signposts outside the remains of an old building at Birdsville.

A desert community. Aboriginal settlement at Warburton Ranges, Western Australia.

The daily rounds. Community Health Sister Jim Craig visiting Aboriginal woman at Oodnadatta.

Lifeline to the outside world. AIM nurse using radio to contact the flying doctor.

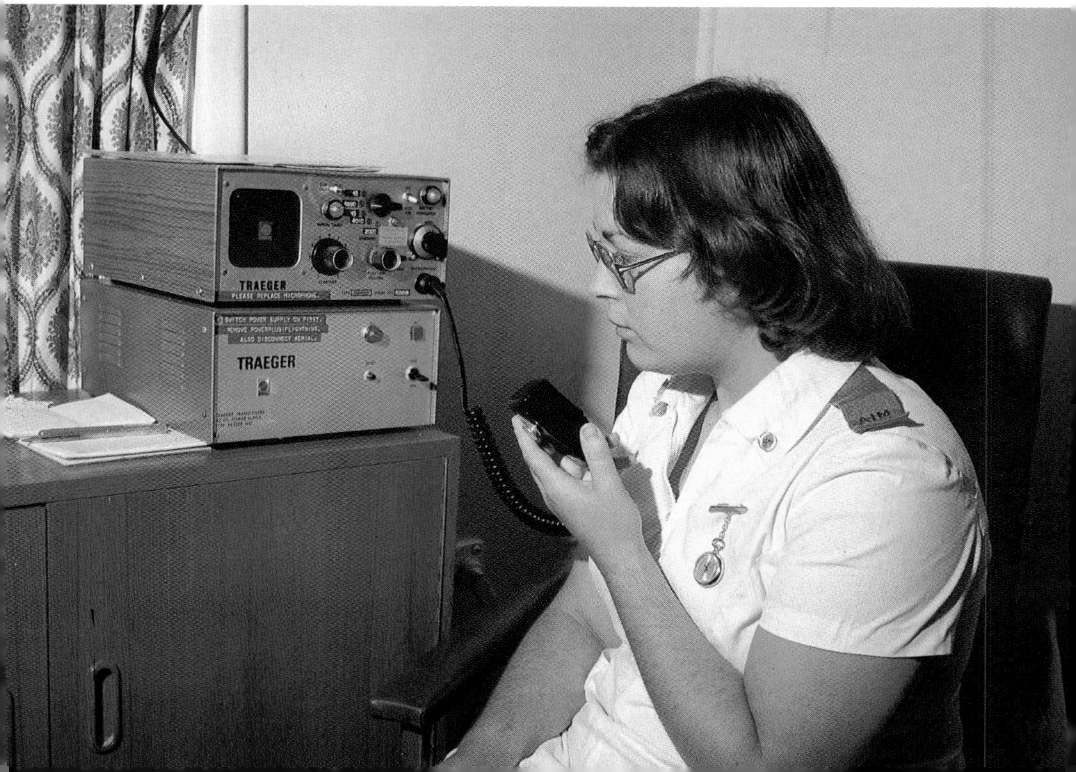

The hospital was at the other end of the main street. Between the Birdsville Pub where we were standing and the hospital, a handful of buildings were scattered. I noticed the crumbled remains of some stone walls across the road from us, obviously the ruins of a building, and recognised the site of Birdsville's first hospital.

'It used to be a hotel,' explained the nurse, noticing my interest. 'It was built in 1886, when Birdsville had three hotels and a cordial factory.'

The Royal Hotel, as it was called, was one of the two hotels which eventually closed when Birdsville fell from the giddy heights of former glories. It remained decrepit and disused until 1923, when Flynn of the Inland was asked to open a hospital in Birdsville. Ever the opportunist, Flynn commandeered the old Royal Hotel as temporary quarters for the first nurses. Thirteen years later, the nurses were still living and working in the temporary accommodation. As I said, the nurses at Birdsville were used to improvising.

Further down the street we came to a squat stone building painted white. It looked like an Indian trading post.

'That's the store,' one of the nurses told me, 'and that's the hospital on the other side of the street.'

I looked across to a long building clad in galvanised iron. It had a steep pitched roof. A windmill towered above it at one end. There were two large water tanks on stands at the other end. If I hadn't known this was cattle country I'd have sworn it was a shearing shed.

We crossed the road and entered into a wide enclosed verandah. Two elderly Aboriginal women were sitting on steel folding chairs, 'up for their daily medication', one of the nurses explained. We dumped our bags and walked through to the kitchen. Like the Oodnadatta Hospital, the kitchen was spacious and the centrepiece was a large table. We all sat down and the mandatory cup of tea was served.

Minutes later, a doorbell rang. Nobody got up to answer it. Then the kitchen door opened and a woman walked in, sat down at the table and pulled out a packet of cigarettes. I had seen her not long before, when we arrived at the airstrip and were unloading our plane.

Another plane had landed just after us. It was the weekly mail plane from the south. The woman had been waiting on the strip with her husband and a girl who was obviously their daughter. There were two suitcases. I learned later that the girl was returning to a boarding school down south. She stood with the barely controlled impatience of someone who wanted to get the waiting over and done with. While her father bent down and fiddled with the straps of the suitcases, the mother took some money from her purse and handed it to the girl. She took it without looking at what it was and, as if half-ashamed, quickly transferred it to her purse. The mother continued to fuss about something. The girl continued to look in the other direction.

I guessed that the parents and their daughter were expressing the pain of parting in the way of people who find it difficult to give expression to their feelings. It was a relief to all of them when the girl finally moved away to board the plane.

The mother who had just walked into the hospital kitchen began to unload all the feelings she had not been able to express at the airstrip. From the moment she sat down, she talked a blue streak about everything and anything. She looked at the smoke curling up from her cigarette as if it came from a funeral pyre.

'Reg stopped off at the pub,' she said, explaining the absence of her husband. 'We're trucking out some cattle and he wants to fix up the time.'

We talked about her daughter's schooling and how expensive the fees were.

'We might not be able to afford it next year, if the drought continues,' she said. 'I don't know what we'll do then.'

The doorbell rang again and we heard heavy footsteps crossing the hall. Then the husband walked into the kitchen. 'Better be on the road, Mum,' he said. 'Got a few things to do when we get back.'

The woman stubbed out her cigarette and rose reluctantly from the table. I walked with them to the door.

'I hope your daughter does well at school this year,' I remarked.

'She's a smart kid,' said the father with some pride. 'Don't know what she'll do when she finishes next year. She wants to come back and work with me. But I dunno, it's no place for a girl, although she'd run rings around most men when it comes to work.'

They drove off in their truck, both sitting upright, with their eyes fixed on the road ahead. I walked back to the kitchen.

'That woman is not well,' said one of the nurses. 'And she misses her daughter like anything when she goes back to school. If we had a vehicle here,' she went on, turning the full force of her concern on me, 'I would drive out and visit her next week. Just a call, of course, on the way to one or two other stations. But I bet that within an hour, I'd have out of her what the trouble was.'

The doorbell rang again. No-one came into the kitchen and after a while, one of the nurses said, 'That will be a patient', got up and went out.

I asked the other nurse what needed to be done around the hospital. Without a word, she handed me a piece of paper. It contained a long list of repairs and replacements that were urgent. I picked one of them at random.

'This blocked sink,' I enquired. 'What's the problem?'

'It's the one in the treatment room,' said the nurse. 'The flying doctor was putting a patient's arm in plaster the other day. He forgot what he was doing and poured what was left down the sink. It's probably set like concrete. But we hear there's a plumber in town. You might like to find him and see if he can help.'

I went down to the pub, where the movements of a fly on the Birdsville Track would be available if you wanted to know.

'He's down at the caravan park,' said the barman, in answer to my query, 'He was driving north for some fishing in the gulf and hit a 'roo.'

I went down to the caravan park, which was a small piece of the Simpson Desert with a fence around it. The plumber was only too happy to make himself useful. He called his mate and we walked over to the hospital.

There was a large cellar under the hospital kitchen which served as a cool room, and the outlet pipes from the sink had been fixed to its ceiling. With some difficulty, we managed to unscrew the offending pipe. The plumber fiddled for a while but without success.

'Go and get my gun,' said the plumber to his mate.

'Who are you going to shoot?' I asked nervously.

The plumber laughed. 'It's an old trick,' he explained. 'I use soft-nose bullets. When they hit the obstruction, it explodes.'

His mate returned with the gun.

'Go out and watch the end of the pipe,' ordered the plumber.

His mate disappeared again up the cellar steps. I had already noted that the outlet pipes from the hospital were not connected to any system. The country round about took every drop of liquid it could lay its hands on, so the pipes simply emptied onto the ground.

The plumber loaded his gun, placed the muzzle at the entry to the pipe and fired twice in rapid succession. The noise in the confined space of the cellar was deafening. The plumber waited for a second and fired once more into the pipe. I waited until my ears stopped ringing.

The plumber's mate came stumbling down the cellar stairs, white-faced and shaking.

'You nearly killed me,' he shouted. 'I heard the first two shots and bent down to see if the water was flowing. The third shot nearly blew me head off!'

'Well, she's clear then,' said the plumber, with some satisfaction.

Improvisation of this kind was a way of life in the outback. But you couldn't always depend on a 'roo to wreck the radiator of a passing plumber. Sometimes it might be weeks or months before a tradesman came along and a repair could be effected.

What worried me more was the improvisation needed to repair a human being. On a subsequent trip to Birdsville, I learned how serious that could be.

The occasion was the Birdsville Races. There was a time when the event was a friendly gathering of people from the town and the district. Sometimes, a few visitors came up for the occasion. Then, one year, a prime minister called Malcolm Fraser decided to go to the Birdsville Races as a goodwill gesture. Somehow his visit acted as the catalyst for a mass invasion. The Birdsville Races have never been the same since. Thousands of people now descend on the hapless community every year.

The first invasion created all kinds of problems for the nurses at the hospital. Most took the form of self-inflicted injuries. People arrived at the Birdsville airstrip, stepped off the plane, stopped off at the pub and went no further until something brought them to the hospital for treatment. At the end of the races, the nurses were exhausted.

The following year we were better prepared. The flying doctor came up from Charleville and brought one or two extra pairs of hands. I persuaded a friend to fly me over from Alice Springs in a little plane. We arrived over Birdsville to find the sky filled with aircraft. Trying to land was like trying to lie down in the middle of a cattle stampede. Somehow we landed safely and walked to the hospital, where the nurses were already doing a roaring business.

We pitched in, stripped beds, washed sheets, cooked meals, kept the drunks away and relieved the grateful patients of conscience money in the form of donations to the hospital.

The races lasted three days and on the final afternoon, the flying doctor at that time, the legendary Tim O'Leary, stormed out of the hospital vowing that this was the last bloody race meeting he would attend. We went down to the strip and waved him off, then trooped back to the hospital and collapsed over the kitchen table.

'Who'll make the tea?' asked a weary nurse.

Before anyone could answer, the doorbell rang but no-one came in. Obviously it was not a local. I went out to see who was there. A group of bikies stood outside, looking uncertainly at me. One of them, a large fellow with a beard, was supporting his arm and seemed to be in considerable pain.

'He hit a pothole filled with bulldust,' said one of his mates. 'He went clean over the handlebars. We think he's busted his shoulder.'

I went to get the nurses. Their weariness dropped from them like a discarded jacket and in a second they had the bikie on the table in the clinic.

'We think his shoulder is fractured,' one of them said, 'but we can't be sure. We'll need to evacuate him to Charleville.'

One of the nurses went into the radio room to contact Charleville base hospital. She came back after a while with a look of concern on her face.

'Tim's plane was on its way back to Charleville when it was diverted to an emergency,' she reported. She thought for a moment and then said, 'We'll try Mount Isa.'

Birdsville is in the middle of a large empty hole in the outback. There are no major medical facilities for 800 kilometres in any direction. Any help would have to come from Charleville to the east, Mount Isa to the north, Alice Springs to the west or Broken Hill to the south.

The nurse radioed Mount Isa and found that their aircraft was on a mercy mission up towards the Gulf of Carpentaria. Alice and Broken Hill also reported that their planes were out on jobs. It was one of those days.

The nurses were in a quandary. The bikie was in great pain and the longer his arm was left unattended, the worse it would become. He needed urgent treatment. One of the nurses went back to have another radio consultation with the base hospital at Charleville. When she returned her face was even longer.

'Charleville has had another emergency call,' she reported, telling us that someone had slammed a car door on her hand.

We all looked at each other in dismay. Two emergencies and no plane. To add to the tension, the sun had done its bit for the day and was showing signs of withdrawal symptoms. If something were to be done, it would need to be done quickly or darkness would put an end to the day.

'I haven't done any night flying for about ten years, but if it's any use, I'll give it a go.'

The voice was so quiet that no-one paid any immediate attention. Then I turned to my friend Peter, who had flown me over from Alice Springs.'

'Someone will have to authorise an emergency flight,' he went on. 'I think it might be permitted under the circumstances.'

It took the nurses about two seconds to make up their minds. They rushed to the radio and contacted Charleville again. They were back quickly.

'Charleville thinks you'd better go to Mount Isa. They will fix up the emergency flight authority.'

The nurses then contacted the station to the north and told them that their patient would be picked up and flown to Mount Isa.

'Be sure you get here before dark,' came the reply. 'The strip's not all that good.'

One of the nurses would travel with Peter, to treat the woman when they reached the station and to keep an eye on the bikie, who was heavily sedated by this time and practically asleep.

We borrowed a vehicle and drove him to the airport. With some difficulty we manoeuvred him into the back seat of the tiny plane. Peter and the nurse climbed aboard.

'Pray for a bright moon,' Peter shouted as the engine blasted into life.

We saw them off and then returned to the hospital to wait. The day died suddenly and I hoped that by then the plane had reached the station, picked up the injured woman and taken off again. A few minutes later we received a radio message from the station reporting that this had taken place. But the little plane was not fast and it would be some time before we learned whether they had made it to Mount Isa.

It was almost midnight before we received a message informing us that 'Your plane has arrived safely. Patients, pilot and nurse all safely tucked up in bed.'

'Well, if they have gone to bed,' said the nurse who had remained, 'so can I.'

Next morning most of Birdsville dropped in, 'just to see how things went last night'. Everyone knew what had happened.

The other topic of conversation around the kitchen table was the wonderful result from the races. I had forgotten about them in the excitement, but the people of Birdsville hadn't. Apparently the result had been record-breaking, financially speaking, which meant that the hospital and the Flying Doctor Service would each receive a big donation. For the townspeople that seemed to make it all worthwhile.

Around 1928, when the nurses were first setting up a hospital in the old Royal Hotel, a famous Australian, Dr C.T. Madigan, was surveying the country around Birdsville. He made an aerial survey of the Simpson Desert and gave it the name. In one of his reports he wrote a brief description of the town of Birdsville. 'A barren and wretched town,' he called it, 'with generous and big-hearted people.'

Anyone who sat around the kitchen table at the Birdsville hospital for a few days would quickly recognise that description, although it was a weary group who sat at the table the morning after the races and the mercy flight. Suddenly the nurse leaped to her feet and said, 'Crikey, the Sched!'

She rushed into the radio room and switched on the transceiver. When it was warm, she began a ritual which had been performed by nurses at Birdsville since the time of the pedal radio. It is called 'the Sched'—short for 'Schedule'.

Twice a day, at 8 a.m. and 6 p.m., the nurses call in every homestead in the surrounding countryside. One by one, each of the stations responds. The Sched is held primarily to provide the homesteads with an opportunity to report any medical problems and receive advice from the nurses. But a bit of friendly chatter always creeps in. There is talk about the weather, requests for the nurses to pick up a parcel from the mail plane and hold it until the people come into town and a bit of friendly gossip about who is visiting them.

'It's a wonderful system,' I remarked to the nurse when she had finished the Sched for the morning. 'Just like seeing your neighbour over the fence or across the street. It helps to know they are there.'

'But it's not enough,' said the nurse. 'What would you do if you were one of those women and you suddenly discovered what you thought was a lump on your breast? Or if you were a man and you found it gave you hell to pass water? Would you want to broadcast that over the Sched for everyone to hear?'

'I don't suppose I would.'

'Then you can see why we want wheels,' she persisted. 'Women won't talk about these things over the radio and bush women are particularly reticent. They are used to putting up with things. But if we happen to drop in and see them, even if it means travelling over some of these rotten tracks for hundreds of kilometres in the middle of summer. . .'

She paused for a moment, as if visualising the scene.

'You're probably right,' I acknowledged.

'I know I'm right.'

We sat in silence for a while and sipped our early morning coffee. Then the doorbell rang. Another day was beginning in Birdsville.

I walked out the back door of the hospital and looked across the barren country which cast a forbidding barrier around the little town. The road which ran past the hospital curled around and dipped down to the causeway, where it crossed the Diamantina River. It rose up again and continued along until it parted company with itself. One track went north to Bedourie and ultimately through to Mount Isa. The other branch headed south and became the Birdsville Track, crossing the hard face of Sturts Stony Desert and going on through Maree to Adelaide.

There, beyond my immediate horizon, lived the people of the outback. They were people of considerable courage, but they were human too. I thought about them and I thought about the nurses at Birdsville, Oodnadatta, Andamooka and elsewhere who didn't want vehicles for joyriding, but to extend the measure of their concern for their people. A set of wheels wasn't much to ask. If John Flynn could put wings on doctors, surely we could put wheels on nurses, for the same reason that Flynn created the Flying Doctor Service: 'To be with people, where people are'.

So when General Motors rang me from Adelaide a few weeks later and asked what I wanted to do with the fully reconditioned two millionth Holden, I had the answer ready.

'Drive it to Birdsville,' I told them.

I flew with a friend to Adelaide and picked up the vehicle. It had been given a heart transplant and a lot of cosmetic surgery. Still, I didn't think the designer envisaged the vehicle ending its life on the Birdsville Track.

We drove to Maree and spent the night. Next morning we set off up the track, which the police at Maree told us was in good condition. So it was, until the rains came. We were about halfway, when a thunderous armada of storm clouds came over the horizon and proceeded to deliver an avalanche of rain. The gibber plains over which we were travelling were as hard as iron. But with the volume of water running over them, the track had disappeared out of sight.

When we came into the sand country the ground under our wheels began to sag. We bogged several times, but somehow managed to get out. The rain became heavier and the light began to fail. I knew that if we stopped, we might be stuck for a week. So we kept going, slipping and slithering up and down sandhills, urging the car not to let us down. Towards midnight we saw, through the blinding rain, a dilapidated old signpost on the side of the road. It simply said, 'Birdsville, 14 miles'.

Sometime later we crossed the causeway over the Diamantina River and pulled up outside the hospital. The town was in pitch blackness. Birdsville had gone to bed. We stepped out into the dark wet night, walked over to the hospital and stepped onto the verandah. The hospital was in darkness. I tried the front door and it opened.

'Is any one at home?' I enquired in a voice that hoped to wake someone, but not everyone.

A door opened across the hall and a female figure appeared, wearing a dressing gown and holding a torch—more for protection than illumination, I thought.

'What do you want?' asked the figure.

'We want to deliver a car,' I replied, feeling it was not a convincing response.

A second figure appeared and switched on the lights. The two nurses looked at the wet weary bedraggled characters in front of them. Suddenly, there was a burst of recognition.

Both the nurses rushed to the door and plunged outside, oblivious to the night and to the rain. We followed them at a slightly more sedate pace. It was an extraordinary sight. There at midnight, in the main street of a blackened and drenched Birdsville, two nurses in dressing gowns were dancing around a mud-covered old car that should have been in a museum, shouting 'We've got wheels! We've got wheels!' at the top of their voices.

Well, it was a beginning.

6

A Forgotten Community

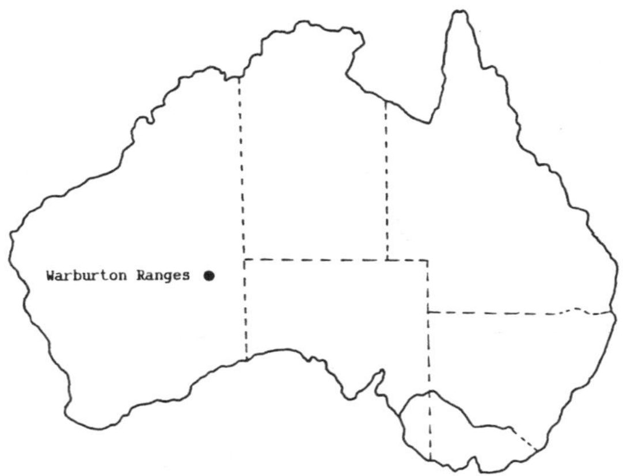

It took me quite a while to find it, and I couldn't do so unaided. I had been poring over a map of Australia looking for a place I had not heard of before. It was somewhere in that great blank space on the map of Australia which covers half the continent and represents desert.

Eventually I gave up and went in search of help.

'Warburton Ranges, Harry,' I said to one of my colleagues. 'Where the heck is Warburton Ranges?'

Harry had been in the outback game for a long time. He didn't even look up, but fished around in his desk and pulled out a much folded and used map. 'It probably isn't on yours,' he replied. 'But it's on this one. If you draw a line from Kalgoorlie to Alice Springs, it's somewhere about the halfway mark. Just a bit north of the line if I remember rightly.'

I went back to my office, unfolded the map and went to work. There was a thick red line running north from Kalgoorlie to a place called Leonora. It meant the road was probably reasonable. Then another red line, much thinner, headed north-west from Leonora towards the Northern Territory border. The condition of that part of the road could be anything. It passed between the Gibson Desert to the north and the Great

Victorian Desert to the south. The thin red line continued across the border, said 'hello' to Uluru and the Olgas as it passed and meandered its way through to Alice Springs.

The distance from Kalgoorlie to Alice Springs appeared to be about 2000 kilometres. Halfway along the line which connected them was a small red speck. Alongside it, in spidery italics, was one word. 'Warburton.'

I looked at it for some time and noted that it was remarkably isolated. So I went back to Harry's office.

'Next question,' I said. 'How do I get there?'

'Ah, now,' he said, smiling and assuming his air of worldly wisdom. 'That will take a bit longer to answer.'

I pulled up a chair and sat down. Harry's explanations were always meticulously detailed.

'It depends on what you want to do. If you want to experience the great Australian outback, you will take a four-wheel drive, either from Kalgoorlie or Alice. On the other hand, if you have any sense, you will fly.'

It was January 1974 and I had just commenced work as superintendent of the Australian Inland Mission. There was a lot of travelling to be done and already the summer had given notice that it would be long, hot and dry.

'I think that on this occasion, I'll be sensible,' I replied.

Harry went on to tell me about Warburton and its background. By 1974 Australia had become uncomfortably aware of the shocking plight of its Aboriginal people. Governments were getting alternatively nervous about backlashes and self-righteous about their desire to redress wrongs. They were particularly concerned about the health of the Aboriginal people. Stories of Aboriginal children dying like flies did not make good reading in the national press, let alone the world scene. Any project which addressed the improvement of Aboriginal health got a good hearing and money seemed to be no problem.

Warburton was an Aboriginal community in Western Australia. As it was remote and difficult to visit, it had suffered from considerable neglect.

The Warburton community had been run by a religious mission for about fifty years. Missionaries first made contact with the Aborigines in the desert and gave them hand-outs of flour and sugar. The spot where they met with the Aborigines was near the Warburton Ranges. Eventually what began as a place for handing out rations, became a settled mission.

For decades, the mission conducted a school, a store and a hospital as well as providing the benefits of their religion. But traditional mission practices were breaking down all over Australia. At Warburton, the school had already been handed over to the government. Now the mission said that it could no longer operate the hospital.

We had operated hospitals in places where there were Aboriginal people since the days of Flynn of the Inland, so I was not surprised when the Western Australian government asked us if we would run the hospital at Warburton. However, since I knew nothing about the place, I said I would go and have a look before answering.

'The best opportunity for flying to Warburton is to ask our friends in the Flying Doctor Service to take you up,' said Harry. 'They have a base at Kalgoorlie and do a clinic run to Warburton every month.'

'I wonder how many Australians realise that there is a place in this country where the nearest doctor lives 800 kilometres away and visits once a month?'

'Maybe that's why they are so healthy,' replied Harry drily. 'The white people I mean,' he added hurriedly. 'For the Aborigines, it's another question.'

We contacted the Flying Doctor base at Kalgoorlie and arranged to go with them on their next visit to Warburton. In due course, I flew to Perth with my business manager and then on to Kalgoorlie. There is probably no other country in the world where you have to fly across the continent and half way back again to get to a place like Warburton. It is literally in the middle of nowhere.

'The news is not good,' said the manager of the Flying Doctor base when we arrived. 'We've had a report that two inches of rain fell at Warburton yesterday and the strip is a washout.'

I looked up at the brilliant blue sky and felt the burning heat of the sun.

'I find that hard to believe,' I replied, knowing that this part of the country had been in the grip of a long drought.

'So do we,' replied the manager. 'The trouble is that radio communication with Warburton is dreadful at the best of times. The hospital up there only has a little transceiver and I think they use the clothesline as an aerial.'

'What will you do?' I asked the manager, hoping I hadn't flown across Australia and half way back again for no purpose.

'We'll go,' he replied. 'Tomorrow morning, first light.'

Next morning we took off in the twin-engine Piper Navajo the flying doctor used for clinic flights and emergency evacuations. The country beneath us was a study in brown, still blanketed in the early morning mist. It must have looked the same since the beginning of time. Not even the rising sun could break its bare monotony. Apart from an occasional dry salt pan, there was nothing to see.

This didn't help the pilot very much. 'They don't have a radio beacon at Warburton,' he informed me. 'So we have to fly visual.'

We both looked down at the anonymous landscape beneath us.

'As you can see,' the pilot continued, 'there's not much help down there either.'

We had flown for several hours when the pilot broke the silence again.

'There's Warburton, straight ahead.'

The horizon was still blurred by the morning mist. Although I strained my eyes, I could see nothing. Then a weak ray of sunshine glinted momentarily on something. As we flew closer, I could make out the iron roofs of a small huddle of buildings. I was looking at the settlement of Warburton. It looked as if it was pretending it wasn't there.

'Where are the Warburton Ranges?' I asked, looking for some prominent landmark. He pointed to a low ripple of hills on the horizon. Colonel Warburton, the explorer, had not exactly discovered a competitor for Mount Everest.

The pilot made a sweep over the town. Everyone looked down anxiously to catch their first sight of the airstrip. If it was awash with water, as had been reported, it was a long way back to Kalgoorlie. But the airstrip was as dry and as bare as an ironing board. Its hard red surface glared back at us and there was not a drop of water in sight anywhere.

'So much for the weather report,' said the pilot with a sigh of relief.

We never did get to the bottom of the mystery of the alleged downpour, but there were several theories. The most imaginative, and therefore the one most widely voiced, was that a small Aboriginal boy had filled the rain gauge in a rather unconventional way. But then, the Aborigines get blamed for everything.

As the Navajo floated down to land, I looked out at the strip ahead. It was lined on both sides with the rusted hulks of abandoned cars.

'The Aborigines buy old bombs in Kalgoorlie,' explained the pilot. 'These are the ones that made it to Warburton. The track up here is littered with those that didn't.'

We taxied to the end of the strip. On closer inspection, the settlement of Warburton was no more impressive than it had been from the air. There were probably twenty houses and a few other buildings, all of a very basic design. A handful of trees huddled around them. They looked as if they were the sole survivors from a stripped desert.

The pilot switched off the engines and the silence of the empty landscape began to filter into the cabin. There was no-one in sight.

Then from the lifeless backdrop of the desert, the figure of a woman emerged and started to walk towards us. She was slight in build, diffident in manner and wore the simple white uniform dress of a nurse. We climbed down from the plane and stretched our limbs as we waited for her to reach us. The Flying Doctor did the introductions.

'This is Phyllis,' he said. 'She runs the hospital.'

'How long have you been here?' I enquired.

'Eight years.'

'And how often do you get out?' I suppose I sounded like someone asking a prisoner about parole.

'I come from Melbourne and I've been home twice in that time. My mother is not getting any younger and I worry about her. It's one of the reasons why I have to leave.'

'What are the others?' I asked, feeling that this woman could tell me more about the place than anyone else.

'It's just that we're not getting anywhere,' she said slowly. 'Things have changed. The mission isn't really interested in the hospital any more. There should be two nurses out here. Sometimes they send another one to work with me, but they don't last long. The isolation and the problems in the community get to them and they leave.'

We had reached the settlement and were trudging along the dusty tracks that ran between the buildings. There seemed to be dogs everywhere, roaming in packs and engaging each other in sporadic warfare. I commented on this to Phyllis, who laughed.

'There are about a thousand of them,' she told me, 'which makes three or four for every member of the community. The Aborigines are very fond of them. They talk about a two dog night or a three dog night, which is like the number of blankets we use. But they're a terrible health problem. I've talked to the council about it and they agree. But they can't bring themselves to do anything about it.'

I asked about the trees, or the lack of them.

'I believe there were a lot more, but they were cut down for firewood. These are all that remain.'

Then she stopped and said, 'This is the hospital.'

I found myself looking at an old iron-clad house. There was a verandah across the front and running down one side. But what riveted my gaze was the roof. It was covered with big rocks.

'What on earth are all those rocks doing?'

'They hold the roofing iron on,' replied Phyllis calmly.

Inside the hospital was dark, dingy and primitive. In one room, the ceiling, from which hung a naked light bulb, looked as if it were about to fall in. There was a bed and an oxygen cylinder standing nearby on the worn linoleum.

'This is the ward where we keep patients,' explained Phyllis. 'We hold the daily clinics on the side verandah. You can see what you are doing out there.'

There was a large room which served as a kitchen and living room. The rooms which the nurses used as bedrooms were predictably gloomy. Hot water was piped from a forty-four-gallon drum outside the kitchen, which was heated by a wood fire.

A couple of mice scurried across the floor.

'We've had a plague of them,' said Phyllis apologetically. 'I don't know which cause me the greater trouble, the mice or the dogs.'

I went back outside and stood in the brightness of the midday sun. Both sky and desert were relentless in their emptiness. The feeling of isolation was oppressive. We were 800 kilometres away from any form of Western civilisation. In particular, we were that far away from any reasonable medical facility, and the community of Warburton was dependent on this decrepit building and the dedication of this woman.

'I marvel you haven't given up years ago,' I told Phyllis.

'I don't begrudge the time I've spent here,' she replied. 'I've come to accept the isolation and the lack of the things I took for granted back home. And I've come to love the people. Despite the difficulties they cause me sometimes. When I get irritated or angry, as I sometimes do, I remember that it's the people here that are the sufferers in the long run, and that stops me from indulging in self-pity.'

I had to remind myself that this was twentieth-century Australia and the great majority of people enjoyed medical services a light years beyond what was available here. Whatever the efforts of Phyllis and the support of the flying doctor, Warburton was a forgotten community at the far end of a forgotten region. The fact that it was an Aboriginal community with appalling health problems only added to the insult.

Phyllis took me to see the school. The young headmaster, who was one of the two teachers, was obviously very angry.

'I don't mind you getting a new hospital,' he said, referring to the government's promise to build one as quickly as possible. 'But as you can see, this place is a disgrace as well.'

I asked him what were the major problems he faced in his work, apart from the poor facilities.

'It's the lack of motivation. The young people finish school and there's nothing for them to do. That's why we have so much disturbance around the place. Petrol sniffing is particularly bad. Sometimes the elders send the young boys who get into mischief down to Kalgoorlie or even to Perth. But they come back and it starts all over again.'

I met some of the Aborigines who were members of the council. They were quiet and polite and said they would be pleased if we would take over the hospital. They spoke very highly of 'Sister', meaning Phyllis, but I came away from that encounter with the feeling that I had not discovered what the Aborigines really felt about things. Six years later, when I visited Warburton for the last time, I still had that feeling.

What I did feel, however, was that Warburton was not a happy community.

'The people are a bit ambivalent,' said the community adviser to whom I spoke. 'As you know, there are four tribes and the land belongs to one of them. Whenever there is trouble, the land owners want to push the others off. The other tribes say they want to go back to their homelands, such as Blackstone and Wingelina, and they do from time to time. But then they come back here again for a while.'

Later in the afternoon, the Flying Doctor pilot indicated that it was time to leave. He was anxious to get back to Kalgoorlie before dark. We said our goodbyes. Phyllis had promised that, if we decided to take over the hospital, she would stay on to help the nurses get established. I had a strong feeling she would not be easy to replace.

The flight back to Kalgoorlie was a fitting conclusion to what had been a rather eventful day. We were bringing back a couple of extra passengers and the pilot had to fly the Navajo at a lower level. The hot turbulent air of the late afternoon caused the plane to bucket around the sky.

After a while, a plaintive voice enquired. 'Does any one know where the sick bags are?'

While I was rummaging around in the cabin lockers, trying to find them, another voice said, 'When you find them, I'll have one too. Just in case.'

'Make that three!' added a third voice, trying to sound jovial.

Air sickness is the most infectious disease I know and before long I was handing out sick bags in every direction and the supply was rapidly running out. Desperately, I searched around in the locker and discovered some sealed plastic bags which contained sterilised instruments. I made a management decision that we were more likely to have to deal with air sickness than perform an operation, tore open the bags and announced that this was definitely the last available.

Up in the cockpit, the pilot and the doctor who was sitting alongside him, had both asked for bags. I was feeling squeamish, but was determined to hang on until we landed. Then I heard the pilot say to the doctor, in an urgent voice, 'Would you hold the bag for me please?'

All that came in return was the groan of the doctor filling his own bag.

It was too much. I succumbed!

It was a sick and sorry bunch of people who fell out of the plane when it landed at Kalgoorlie. The late afternoon sun was not lying down without a fight and we hastened across to sit in the shadow of the hangar wall. I found a tap and put my head under it. My business manager, Doug, who had flown with me, came and sat down alongside. There was a long silence as we tried to recoup our strength.

'Doug,' I said finally, 'what are we going to do about Warburton?'

There was another silence, and then Doug said slowly, 'I don't know, Max, but I do know we can't turn our backs on it.'

A few months later we had agreed to manage the hospital and I was on my way back to Warburton. The government had promised that a new building would be started as quickly as possible. We had also been able to recruit four nurses to replace Phyllis. Two of them were to concentrate on improving the health of the community and the other two would staff the hospital.

The four nurses set out from Perth with one of our male staff members. They were driving two Land Rovers and a sedan, as well as towing a large caravan which would be used to provide accommodation until the new hospital was built. At the last minute the party was joined by a journalist from a national women's magazine who wanted to do a story on the women who worked in the outback.

The party reached Kalgoorlie without incident. Then they began the long haul over the track to Warburton. Somewhere along the way the journalist persuaded the driver of the Land Rover which was towing the caravan, to let her take a turn at the wheel, 'just so I can have a ''hands-on'' experience'.

The Warburton track is notorious for its sand drifts. When you have been driving for a time and the sun creates a lazy euphoria, it is not hard to let the vehicle get into a drift. Unfortunately when the journalist got into a drift, she overcorrected on the steering. The Land Rover swung wildly, tipped sideways and the next moment was upside down. The two occupants were also upside down, hanging from their seat belts, while the caravan they were towing broke its coupling and finished up on top of the Land Rover.

When the nurses arrived on the scene minutes later, they were appalled to discover one of their precious vehicles and their future temporary home stacked up in a heap of wreckage. They rushed to investigate what had happened to the occupants, who were not seriously injured. Their feelings about the ineptness of the journalist were remarkably restrained.

Oblivious of all the drama that was happening on the ground, I was flying out to Warburton in one of our own planes, with a padre called Gordon. We planned to be there before the road party and make preparations for their arrival. Being mates, we didn't have to make polite conversation and simply sat in silence, watching the unchangeable desert unfold beneath us. The aircraft radio was turned on as it always is, but turned down so that the static was reduced to a soft burble.

Suddenly, the radio crackled into life.

'All stations!' announced the voice on the radio. 'Accident reported on the Warburton track. Extent of damage or injury unknown at this stage. It is understood that there were nurses in the party.'

Gordon and I stared at each other in dismay. 'That's done it!' exclaimed Gordon. 'We haven't even started at Warburton and we've killed the nurses.'

We flew on to Warburton, landed and hastily commandeered a radio. The land party was about 300 kilometres out. Fortunately, their radio signals had been picked up by the people at a cattle station in the region, who had gone to their help. They had been able to repair the caravan so it could be towed to Warburton, but the wrecked Land Rover would have to be taken back to Kalgoorlie.

The party duly arrived, weary, shaken and dirty. We jacked up the caravan on blocks while the nurses had a well-deserved shower. Afterwards, they had a long hard look at the old hospital. Three of the four nurses were very experienced in working in the outback and I asked them what Gordon and I could do before leaving them to sort things out.

'You're not leaving before you get rid of those mice,' they replied. 'We can cope with everything else.'

Using more methods of assassination than a terrorist squad, Gordon and I killed about 300 mice before we were allowed to leave.

'They seem a competent mob,' said Gordon, as we flew out from Warburton. 'They've had enough drama with that accident to last them for the next twelve months.'

It was a nice thought, but it was not long before I was once again on a plane flying out to Warburton. Communication with the isolated community was still limited to radio contact with Kalgoorlie and so the messages we received were often brief and cryptic. The message I received from the nurses through the Flying Doctor was simply, 'Trouble. Please come.'

By this time there was a regular weekly flight from Perth to Warburton, which stopped at Kalgoorlie on the way. The plane was a small twin-engine Beechcraft Baron owned and operated by a Dutchman called Jan. He was one of the most versatile pilots with whom I have flown, and he had to be, considering the passengers and freight he flew. They ranged from government officials, to cages filled with chooks, to a corpse in a coffin. Jan once remarked to me that at times he couldn't tell the difference.

On this occasion I shared the cabin with a huge dog in a wire cage. It had been sedated for the trip, but unfortunately they couldn't sedate its smell. It was a long trip from Perth and Jan and I were relieved when the plane finally touched down at Warburton.

'They've got enough bloody dogs at Warburton without bringing this one,' snarled Jan.

'Whose is it?'

'It's the family pet of one of the white community workers. What will happen when it sees the other thousand up here, God only knows.'

As the plane swooped over the settlement I could see that the builders had started on the new hospital. But I couldn't see anyone working on the job. I had a sense of foreboding.

The nurses quickly filled me in on the situation. Alcohol was forbidden in the Warburton community. The building team who had come out to erect the hospital were resentful of the fact that they were 800 kilometres from nowhere, with nothing to do for relaxation and even deprived of the simple enjoyment of a drink.

To make matters worse, the Aborigines had come to the builders with the complaint that a trench digger, cutting out a trench to lay the sewerage pipes for the hospital, had cut across the trail of a sacred snake. The Aborigines were demanding compensation, which the builders had refused to make. It was an angry stand-off before a satisfactory solution was negotiated.

Things seemed to be settling down when the builders went off for their Christmas break, but they arrived back to discover that the building had been vandalised and some of their equipment stolen. The enraged builders stormed out of the community, voting never to return. Negotiating a peaceful settlement when the two antagonists were 2000 kilometres apart in distance and even further in culture was not easy.

The nurses, who had been working under impossible conditions in the old hospital, had been prepared to put up with it in the belief that they would shortly be able to move into the new hospital. With each succeeding dispute they saw that possibility disappearing into the unreachable future.

I could only sympathise with them and try to expedite the resolution of the dispute. Eventually the builders came back and the completion of the hospital seemed to be realisable.

Cautiously, we began to plan an official opening of the hospital and invited the Federal Minister for Aboriginal Affairs to do the honours. I was keen to get him out there to see this forgotten community which had been neglected for too long. The minister accepted the invitation, but two weeks later, I received a phone call from Canberra.

'Regarding this opening of the hospital at Warburton,' said the minister's secretary. 'We seem to have a problem.'

'That's not unusual,' I replied. 'What is it this time?'

'We've received a message from the community saying that they were not properly consulted. They say that if the minister arrives by plane to perform the opening, they will not let him into the community. Incidentally,' the secretary went on, 'how do you people communicate with the community? We get vague messages from our office in Perth, but you can't be sure if they are accurate and when they were actually sent.'

I explained to him the eccentricities of the radio system and offered to help.

'I'd be glad if you could,' said the secretary. 'The minister is in a bit of a spot. The state government reckons the Aborigines are playing politics and insists that the opening go ahead. But the minister can't afford to go out there and get the big heave ho.'

We made several attempts to negotiate, but the Aborigines were adamant. They would not let the minister in to perform the opening. Several weeks went by and the stand-off continued.

Then one day, out of the blue, I received a letter from the nurses at Warburton. I opened it with apprehension, because every message by now seemed to be filled with bad news, but I read on with growing astonishment and finally with amusement and relief. The letter read roughly as follows:

'Well, we suppose you are wondering how the opening of the hospital went.' (That was the first shock. I didn't know there had been an opening.) 'We held it at eight in the morning and it was a nice sunny day. The whole community turned up and the children from the school sang some songs. Jenny [one of the nurses] gave a short speech and then Terry Robinson [the Chairman of the Aboriginal Council] said a few words and cut the red crepe streamer we had stretched across the verandah. Then everyone went in to inspect their new hospital. There were a lot of oohs and ahhs. Afterwards we all went outside and sat on the ground and had a party of damper and jam and tea.'

I finished reading the letter and sat back reflectively. Sometime, in the days ahead, a bureaucrat in Canberra would come across a file marked, 'Warburton Hospital : Official Opening : Pending', and scratch his head and wonder what it was all about.

The nurses at Warburton at the time of the opening were a very experienced and competent team. They seemed to enjoy the respect of the community. Once again I hoped that, this time, things would settle down for a while. Once again I was wrong. It was not long before I was back on the plane to Warburton, and the tension in the community was apparent from the moment I got off the plane.

You can generally tell when an Aborigine is angry. He walks with a stiff, angular

Outpatient clinic in the bush. Sister Jean Williams works on the verandah at the old hospital, Warburton Ranges.

The 'unofficial' official opening of the new Warburton Ranges hospital. Sister Jenny Meyenn cuts the ribbon.

New day dawning. An Aboriginal woman and her baby delivered at Warburton Ranges hospital.

Twentieth century nomads. Construction workers live in caravans at a new mining site, Middlemount, Bowen Basin.

gait, which is in contrast to his usual loose-limbed fluid movement. When I walked up to the settlement from the airstrip, there were Aborigines stalking along the streets carrying spears and other weapons. There were sharp heated exchanges in their language. You could feel the storm building up.

I went into the hospital. The nurses told me that there had been a number of outbreaks over the past few weeks. While we were talking, the sound of a great commotion came from the street outside. I went with Murray, our handyman, to see what was happening. It was an amazing sight. There were over a hundred Aborigines engaged in pitched battle. They were hurling spears at each other and wielding iron bars and other weapons. There was a lot of shouting. I wondered how you could stop such an outburst of raw passion.

Suddenly a spear came hurtling through the air and struck the ground beside me. I bent down to pick it up.

'Don't touch it!' warned Murray.

It was too late. Someone thumped me in the side of the body and wrenched the spear away.

'We'd better go back inside,' said Murray. 'It'll be over soon and then the nurses will have to clean up the mess.'

We went back into the hospital. After a while the noise subsided. I looked out and there was no-one in sight. I went back inside again. Then the bell on the clinic door rang. The nurses gave a sigh of resignation and went to deal with the casualties. I went with them. Despite the ferocity displayed during the fighting, the amount of serious injury done to people was surprisingly small. Nevertheless, the nurses were kept busy stitching up cuts and dressing wounds.

'The Aborigines seem to be able to direct their weapons with considerable accuracy,' I commented, when the nurses had finished. 'I was expecting to have several fatalities, at least.'

'We had a case recently,' replied one of the nurses, 'where a man was brought in from one of the outstations. He had been wounded in a fight. I was pushing him along in a wheelchair when a group of Aborigines suddenly appeared through the front door. They were carrying spears.' She stopped and took a deep breath as if reliving the whole situation. 'Without a word being spoken, they walked up to the man in the wheelchair and speared him in the thigh. Then they turned and walked out again. I gather it was a payback.'

'Imagine the uproar that would create in the Royal Melbourne Hospital,' I commented.

The day after the fight, I went with one of the nurses to visit a homeland community in a place called Blackstone Ranges. It was about 200 kilometres further inland, close to the Northern Territory border. The track was very rough, with extensive sections of rock outcrop and some steep sandhills. The old Land Rover chuffed and chugged its way along and displayed its displeasure at some of the sandhills it was asked to climb.

Finally, the nurse pointed to a windmill on the horizon. 'That will be the Blackstone community,' she said.

If ever I needed to be assured that the Aboriginal homeland movement was genuine, Blackstone Ranges convinced me. There was no reason why people would come and live in this barren featureless spot unless it had a special meaning for them. A bore had been sunk and some trees planted. An airstrip had been graded and there were a few simple dwellings. There was also a caravan.

'That's the clinic,' said the nurse. 'A young Aboriginal woman runs it and the flying doctor comes in once a month when he is visiting us at Warburton. We pop out to see her when we can.'

I thought about the 200 kilometres of rough track over which we had just driven and the expression, 'popping out', didn't quite fit.

We chatted with the Aboriginal woman in the clinic.

'I give out medicine,' she explained. 'Sometimes it's a bit hard if one of my relatives comes and demands to be given something which I don't think they should have.'

'What do you do?' I asked.

'I slam the door,' she said firmly.

Later we returned to our vehicle. I looked back at the simple settlement we were leaving. 'If the future of the Aboriginal people depends on the development of places like this,' I commented, 'it's going to be a long haul.'

'It's what the people want,' replied the nurse quietly. 'And I suppose, in the end, that's what matters.'

The nurse took the wheel on the return journey. 'There's one sandhill that I always have trouble getting over,' she said. 'I generally finish up getting bogged on the top. It's my big challenge and today I'm determined to get over it.'

We reached the unsuspecting sandhill and stopped just short of it while we discussed tactics. Then the nurse put the Land Rover in gear and went for the hill as if she were leading the Charge of the Light Brigade. It had a steep pitch and the ridge at the top was sharp. The nurse urged the old vehicle along and made it to the top. There was a heart-stopping moment when it looked as if the Land Rover might stall. Then she gave it an extra thrust of the accelerator and we were over. The nurse stood tall in the saddle as if she had just completed the water jump of the Grand National Steeple.

Back at the hospital we were greeted with some anxious faced nurses. 'We have a young woman in labour,' one of them told us. 'We think the baby is inverted.'

'Can we evacuate her?' asked the nurse who had just arrived.

'I've contacted the flying doctor at Kalgoorlie,' her colleague said. 'But it's too late in the day. He couldn't get here until tomorrow morning.'

'What can you do?' I asked the nurses.

'We can't do a caesarean,' replied one of them, 'so we'll have to hope for a normal birth. We'll sedate the mother because it's going to be a long job. But if we give her too much sedation, we might sedate the baby and that won't help.'

The next eight hours were tense. From time to time the nurses radioed the doctor, 800 kilometers away in Kalgoorlie. The reception was not good and sometimes it was difficult to understand what he was suggesting. Once the nurse had to ask the doctor to repeat several times the dosage of the sedation. She had almost given up asking when another voice cut in.

'He said, give her ten mils.'

It was obvious that there were other people in the region who were listening in to the drama, though goodness knows where.

Around midnight, I fell asleep. The noise which woke me sounded as if someone had won the lottery. After all the anxiety and waiting, the baby had decided to enter the world on its own terms. I'm sure the doctor at Kalgoorlie didn't need to be told by radio. The celebration could probably be heard in Perth.

The relationship between the community at Warburton and the hospital continued to vacillate. Sometimes it seemed that things were going very well. I would receive reports from the nurses that they had been out hunting with the women. They were amazed at the accuracy with which the Aboriginal women used shanghais to bring down birds on the wing. Jean, one of the first nurses to work at Warburton, was a keen painter. She discovered that, unlike other Aboriginal communities, the people at Warburton had never painted. So she taught them.

Then the next report would tell of an outbreak of violence and the stoning of the hospital by some of the Aborigines. Generally, the nurses had no idea what brought on the attacks.

The next time I visited Warburton, I drove up the track from Kalgoorlie with a padre called Bill. He let me take a turn at the wheel, but warned of some unusual hazards on the road.

'The last time I drove up here,' said Bill, puffing on his pipe, 'I was going along, minding my own business, when suddenly, a great bull camel came plunging onto the road in front of me. Instead of going straight across, it turned and started running up the road in front of me. I was travelling pretty fast at the time. I applied the brakes, but the vehicle simply skidded in the sand.'

Bill took his pipe out of his mouth and grinned.

'All I could see was this great backside of the camel getting closer and closer and I thought to myself, What a way to go!'

We lapsed into silence for a while as I concentrated on a difficult stretch of sand. Then Bill began to speak again.

'The problem up there,' he said, pointing in the general direction of Warburton with his pipe, 'is a breakdown in authority. The elders have no control over the younger men. We had a case just a while ago. Some of the boys who were petrol sniffers broke into the hospital yard to get petrol out of the Land Rovers. They did a bit of damage.

'The elders called the police at Laverton. But by the time they had driven the 600 Ks up the track, the villains had flown out into the desert somewhere. The police made a half-hearted attempt to go after them, but it was too late. So they gave up and drove the 600 Ks back again to Laverton. It's pretty frustrating for them.'

We were heading up to Warburton because the nurses had been under attack again. When we reached the hospital, they were obviously upset.

'We had a big row the other day,' a nurse told me, 'and after things quietened down, people came into the hospital to be patched up, as usual. Suddenly fighting broke out in the clinic. We tried to restore order, but one of the women attacked us. In the end, we got them all out of the hospital and said that we wouldn't open again until they stopped fighting.'

'What do you think is the real trouble?' I asked.

There was a pause for a while and it was obvious the nurses were not quite sure.

'I think the hospital is partly to blame,' said one of them slowly. 'It's too big and it's almost obscenely modern in comparison with everything else in the community.'

I was forced to agree. The hospital had been built on a plan which was developed for the hospital at Fitzroy Crossing in the Kimberley region. But the Aboriginal population at Fitzroy was about ten times the population at Warburton and it was one of the first buildings in what became a brand new town. It was also built for a cyclonic region. Warburton, which was several thousand kilometres inland, certainly didn't need all that protective armour.

Another nurse smiled and added, 'I think the people throw rocks at it because it's so big they can't miss it.'

'Bill's right when he talks about the lack of authority,' the first nurse continued. 'We go and talk with the elders and they promise to do something about it. It's not their fault. They just don't have the authority in the community they used to have.'

I went back from Warburton with a heavy heart, wondering how much longer the nurses could cope with the tension and violence. I talked with the government and the Aboriginal Affairs people. They were all sympathetic, but no-one seemed to be able to do anything.

The next few months were a nightmare. On Christmas Day 1978 tempers blazed up in the community and there was a big fight. Some of the Aborigines broke into the community store and looted it. The nurses' home was stoned and one of them was attacked.

Two months later, there was another outbreak of fighting which lasted for several days. The hospital was stoned on four occasions. Windows were smashed and the fuse box ripped from the hospital wall. An Aboriginal boy, who was addicted to petrol sniffing, was caught trying to steal a hospital vehicle.

A little later, there was further fighting. When it had finished, the hospital clinic was filled with people waiting to be patched up. Feelings were running high. Fighting broke out in the clinic and one of the nurses was attacked.

Three days later, someone came into the clinic and assaulted a nurse because she would not do what he wanted. The nurse had had enough. She left on the next plane. The two nurses who remained were determined not to give up. But they had been working at Warburton for twelve months under conditions of great stress and personal risk. I was not surprised, therefore, when I received a radioed message from the nurses, tending their resignations.

In a letter which followed, they said they could not continue to work in a situation where there was no respect for them, either personally or professionally. In particular, they pointed to the failure of the Western Australian government to provide them with adequate protection.

I had no argument with their reasons. We had been asking the government for years to locate a police presence at Warburton. Towards the end, the Aboriginal community, which had initially resisted the suggestion, was totally supportive of the request. But it was turned down.

So for the last time, I flew out to Warburton. I met with the Aboriginal council and explained that we could not continue to work in a community where the nurses were subjected to abuse and attack. I said that the government had let us and the community down by not providing adequate protection. The only solution I could see was to make the government accept full responsibility for a situation they had chosen to ignore. We were handing the hospital back to them.

I finished and looked around. There were a few tears in the eyes of the people and the nurses. There were some in mine too. Some of the members of the council spoke of their regard for the nurses and their sorrow that we had to go. But they understood why.

I think that at that moment, we all simply felt helpless.

In the years that followed, I often wondered what more we could have done. The nurses who served at Warburton were magnificent. They accepted the isolation and the harsh environment. They came to love the people, even when there was friction and hostility. I think that when they left and had time to reflect, they probably realised that in the end, it was the Aborigines who suffered.

The truth is that no white person can eventually solve their problems. The hospital building was an example of this. It was a white answer to an Aboriginal problem. I think the happiest Aborigines I saw in that region were those who lived out at the Blackstone Ranges, and that was something no white person could understand.

My departure from Warburton was as dramatic as the first time we had flown out in an overcrowded Flying Doctor plane. This time it was the redoubtable Jan who was flying his Beechcraft Baron.

I spent the last day making final arrangements for the hand over to the government and I was so engrossed in talking with people, that I didn't notice the time. Finally one of the nurses rushed in and said, 'Jan is about to take off!' I went outside and saw the Beechcraft taxiing down to the end of the strip in preparation for take-off. I looked around wildly, wondering what to do. There was a Land Rover parked outside the hospital.

'Give me the keys!' I shouted to the nurse. She jumped in beside me. I drove furiously down to the airstrip. The plane was turning for take-off. Without thinking, I drove the Land Rover straight down the middle of the strip. It roared along, making more noise than speed. I pulled up in the centre of the strip, nose-to-nose with the Beechcraft Baron.

Jan poked his head out of the cockpit window.

'Get that damned vehicle off the strip,' he shouted, over the roar of the engines.

I smiled and waved at him, pretending I could not hear what he was saying. Jan continued to shout and I continued to smile helplessly. Eventually, he throttled down the engines and allowed me on board.

As soon as I climbed in, Jan roared the engines and raced down the strip in great fury. I sat and watched Warburton slip away from beneath me for the last time. After a while Jan cooled down, as he generally did. Then he pointed to the sky up ahead of us. We were still climbing.

'Do you see that?' he demanded.

There was a massive build-up of thunder clouds in front of us. The sky was pitch black and ominous. I nodded.

'Another quarter of an hour on the strip and that storm would have hit us. We probably wouldn't have got off for a week.'

I had gone to Warburton on the first occasion, with the report of a storm that had washed out the airstrip. I was leaving Warburton in similar circumstances. Maybe that was the nature of the place.

I sat back and reflected on some words that wise old John Flynn had once spoken: 'You succeed in this business when you do yourself out of a job'. Perhaps one day I would understand whether that was the case for our work in the hospital at Warburton.

7

Urban Transplants

I blinked in astonishment.

A multi-storeyed brick building reared up on the desert horizon, aloof and isolated from its barren surroundings. It was like the last sentinel of some vanished civilisation.

'What's that?' I asked my travelling companion.

'That's Alcatraz,' replied Jon calmly.

Nine years after I first visited the Pilbara region in 1965, I was returning in the role of superintendent of the Australian Inland Mission. Jon and I were driving from Port Hedland to the new satellite town of South Hedland. It had been built to absorb the flood of workers and their families who had come to the north-west to work in the iron ore industry.

'What is it?' I asked, even more bewildered by the bizarre sight of a high-rise in the desert.

'It's actually an apartment building. You can't see from here, but it's located in South Hedland. Every other building in the town is single-storeyed.'

'Why on earth did they stick up something like that, when there is more land than they could ever use.?'

'Well that's just one of the great mysteries in the planning of South Hedland. We'll go and look at some of the others.'

The new housing developments were laid out in clusters of cells. The houses backed onto ring-roads and faced into cul-de-sacs with walkways.

'The design was intended to encourage people to walk,' said Jon. 'It's based on an English scheme.'

We got out of the car and inspected the cul-de-sacs at close range. The fierce sun of the north-west beat down on us.

'It might be all right in the English countryside,' Jon remarked, 'but strolling down a country lane is not my preferred option when the temperature is roasting you at 45°C.'

I turned and gazed to where the apartment block rose high above the rest of the town.

'I still don't know why you call that skyscraper in the desert Alcatraz,' I persisted.

'We'll go and meet one of the residents. I think you might understand then.'

We pulled up outside the five-storeyed, brown-brick apartment building. It might have been plucked straight out of harbourside Sydney. The ground around it was bare concrete, decorated with battered dustbins and a few garden patches which had long ago succumbed to the weeds. We climbed up the concrete stairwell to the third floor. Jon knocked on one of the doors. It was opened by a man in his thirties. He was wearing thongs, black shorts and a singlet that had seen better days.

'This is Vic,' said Jon, by way of introduction.

'Welcome to Alcatraz,' said Vic with a grin, ushering us into the little apartment.

We sat in a couple of worn armchairs and Vic offered us a drink.

'Excuse the flies,' he said, vainly trying to brush them from the table.

I noticed there were no flywire screens on the window.

'Yeah, well, we told the Housing Commission the flies were bad and asked for some screens,' said Vic.

'What happened?'

'Three months later we received a polite letter. It said that scientific research had proved that flies could only reach the level of the second floor, so that was as high as they had installed the screens.'

'Did you protest?' I asked incredulously.

'It took another three months to persuade one of the officials to come and see that we had flies which wore oxygen masks. He said he could see that we had a problem, but we still haven't got the screens.'

As we left the building, I looked back at its forbidding and uncomprising bleakness and thought about the residents who were imprisoned in frustration. I now knew why they had called it Alcatraz.

South Hedland was an urban transplant in the barren body of the desert. It was one of a number of similar transplants which took place in the Australian outback during the '60s and '70s. Most of them were built to accommodate the work forces of the mining industry.

The character and quality of the new mining towns were generally high and Alcatraz was an extraordinary exception. The mining towns in the Pilbara and elsewhere were built at a time when the Australian economy was on a high. If workers and their families were to be encouraged to go and live in remote areas, then their living conditions needed to be attractive. So most of the towns built were like urban transplants, providing the kind of housing and facilities commonly found in the new suburbs of a city.

'Guess what the average age of the population is up here?'

The question was thrown at me by Don, another colleague working in the Pilbara. We were standing in the shopping centre of Tom Price, one of the first of the mining towns built in the Pilbara.

I looked at the people who were walking past me. Most of them were young couples with one or two little children. Others were single men, also young. The casual clothes they wore and their sun-tanned skins made them look even younger.

'About twenty five,' I said, 'at a rough guess.'

'Wrong!' replied Don triumphantly. 'It's six!'

'Six?' I echoed in disbelief.

'Yes, six. There are hundreds of kids here, hardly out of nappies. And at the other end of the scale, if you're over thirty you're over the hill.'

That simple statistic spoke volumes about the character of many of the new mining towns. They were young towns in every sense of the word. An elderly woman from Melbourne who went to visit the Pilbara in 1975 was astonished to hear a small child say to his mother, 'Look mum, there's a Granny!'

I looked around the shopping centre where Don and I were standing. There wasn't a grandmother in sight.

We left the shopping centre and drove around the town. Ten years earlier I had stood and watched earth moving machines ripping up this terrain that had probably remained untouched since time began. Now there was a town of 4000 people, living in neat suburban homes. Not far from the town's perimeters was an open-cut iron ore mine, with large sophisticated machinery, mining the ore which would find itself eventually in the blast furnaces of the mills in Japan.

We drove down streets lined with brown brick houses, neat lawns and flowering shrubs.

'I could be driving through any one of twenty outer suburbs of Melbourne,' I remarked to Don. 'All I need now is an A.V. Jennings sign to tell me how cheap they are.'

Don grinned. 'Jennings built them,' he said. 'The mining company owns them and I think they cost a little bit more than had they been built in Melbourne. But the rent is ridiculously low and the company provides everything, even the shrubs in the garden.'

We turned into another street. Outside every house was a large cardboard container.

'They're washing machines,' explained Don. 'It's washing machine change-over day for this part of the town. Everyone gets a new washing machine, whether you like it or not.'

I remembered a television ad about washing machines, which was popular in the east at the time.

'Guess whose mum's got a Whirlpool?' I jested.

'That's no joke.' Don pulled up the car. 'Get out and have a look.'

I got out. Every mum had a Whirlpool.

The street along which we were driving came to a dead end. So did civilisation. From this point on, the desert resumed its relentless control of the landscape. Don gave words to what I was feeling.

'The houses are great. Air-conditioning, plenty of space. The school's good and there are plenty of sporting facilities. You've seen the shopping centre. We could do with a bit more variety, but it's not bad.' Then he turned and faced the desert again. 'But this is something the company can't do anything about. In the end you come to the desert and accept the fact that you're living in isolation.'

We climbed back into the car and drove to the kindergarten. We had built a kindergarten in the early days of Tom Price, with the help of the company, and it had been doing a roaring trade ever since. We arrived just as the mothers were collecting their children after the morning's session.

'What's it like being a mum out here?' I asked one of the women who had come to collect her child.

She looked a bit embarrassed at first. Then she gave me a rare response of candour.

'You'll probably think I'm stupid if I tell you I miss my mum,' she replied finally. 'But just at this moment, that's what I was thinking and why I was embarrassed. I thought you had read my mind.

'It's the little things, such as what's happened today. The baby's home sick and I had to come and collect Wayne. Then I have to take him to the dentist. Someone's coming to fix the stove. Back home I'd simply ring mum and ask her to come over. But 4000 kilometres is a bit far to drop over for the afternoon.'

That night I dined with one of the senior officers of the mining company and his wife. Their children were older.

'Yes, I'm one of those who are over the hill,' he laughed. 'The ripe old age of thirty-five. We've really enjoyed living up here for the past six years. But now we've got to face up to the future. There's the children's education, for instance. Either we move back to Perth or we send them away.'

He paused for a moment, as if struggling with a decision.

'If we leave, we lose a good job and a good house. We're spoiled with the rent here you know. We've come to enjoy the fresh air and the open lifestyle.'

After dinner we drove back to Don's house.

'There's a bit more to it than that,' said Don, referring to the mining man's comments. 'Take the air-conditioning for instance. When the temperature runs at 45°C for weeks on end, air-conditioning is terrific. But there are women in this town who stay in their house all day, except when they get into their air-conditioned car and drive to the air-conditioned supermarket. They become shut off from the rest of the community and in this kind of isolation, it's not good for them.'

I left Tom Price the next morning. As Don drove me through the streets, the sun was already boisterously bouncing its beams against the brick walls of the houses. It was then I noticed something I had not noticed before.

Everything seemed to be covered with a fine layer of red dust. It clung to the walls of the building. It grubbied the concrete. It smeared the bodywork of the cars. When I touched the upholstery and fittings inside, they had the feel of fine sandpaper. I could feel it on my skin.

The desert was not content to stay on the outskirts of the town. It was insidiously infiltrating every aspect of it.

Don was a flying padre, so he flew me from the mining town to one of the coastal towns of the Pilbara. Far below us, the road snaked across the desert, stretching from one horizon to the other. There was another line that looked as if it had been ruled across the landscape.

'That's the railway,' he told me. 'It takes the ore from the mine to the coast. The trains are a mile long and carry about 20 000 tons. Two of them had a head-on collision the other day. I saw the result from up here. Imagine, two miles of pile up. There were ore trucks everywhere.'

'It must have taken them weeks to clean it up.'

'The trains were running the next day. They simply built a temporary line around the crash site. There's too much money involved to fiddle around.'

I looked down at the vast empty landscape. The Tom Price township may have looked like an urban transplant, but two miles of train crash is not exactly a common occurrence in the cities.

The town to which Don flew me was called Karratha. It was a planned, functional place which seemed to run as smoothly as some of the mining operations I had witnessed.

'It looks like a company town to me,' I remarked to Gray Birch, who lived there. 'Mass-produced houses, all the men wearing hard hats.' I looked down at my trousers, covered in the all-pervasive red film. 'Even the dust is the same.'

Gray was a man with a shrewd mind and a dry humour.

'Karratha is different,' he replied. 'It's an open town. The mining companies are here in strength but they don't control it.'

We had driven away from the town proper and were passing a number of small factories and steel sheds.

'This is the light industrial area,' said Gray. 'The buildings belong to the contractors and suppliers who work for the mining companies.'

Suddenly, he swung the car into the gateway of a property surrounded by a high cyclone fence. Vans were parked close together and the whole place looked dilapidated.

'What's a caravan park doing here?' I asked. 'Who would want to holiday in a place like this?'

'The people who live here aren't on holidays. They're up here to work?'

'But why live in a caravan?'

'They have no choice. In the company towns, you work for the company and get a company house. Once you leave the company, you leave the house. They don't like people just hanging around, so they bar caravans. But in an open town there are a lot of people who have to find their own accommodation.

'Come on,' he said, getting back into his car. 'I'll take you to see Stella. She can explain it better than I can.'

Stella was a middle-aged woman. She had an air of quiet determination. Her husband had a senior job in education and they lived in a pleasant house.

'Al's busy with his work and there are not many jobs for women in a town such as this,' she explained. 'So I told Gray I was looking for something useful to do. He introduced me to the caravan people.'

'Why would people want to live in a caravan in this kind of country and climate?' I demanded.

'They have no choice. The competition for houses is fierce and the rent, if you can get one, is horrific. So it's a caravan or nothing.'

'What's it like for them, living in a caravan with kids?'

'It's rotten,' said Stella with some feeling. 'Take that park you visited. The drains in the ablution block were so badly blocked that from four o'clock in the afternoon until midnight they were twenty centimetres deep in water. You can imagine what it was like. It was absolutely foul and a breeding ground for disease.'

'When the fellows come home from work or the kids from school,' added Gray, 'the first thing they want is a shower. It's the one way to bring back your sanity at the end of the day.'

'But didn't they complain?' I asked.

'Yes, they did.' responded Stella. 'Someone got up a petition and everyone signed it. Then they presented it to the management.'

'Did that do the trick?' I asked.

Stella looked fierce. 'The next day, the management sent out word that unless the petition was torn up, the first ten names on it would be out of the park in twenty-four hours.'

'They can do it, you know,' added Gray. 'Under the law, people aren't suppose to live permanently in a caravan park.'

'So what happened?'

'They tore up the petition,' said Stella. 'I heard about it a few days later and decided to do something.'

She went on to describe how she had approached the shire council and told them of the health hazard that existed in the caravan camp. Someone at the shire office made a note of her complaint, but nothing happened.

'So I went to see the shire president and said that I was going to make a public issue out of it. That seemed to do the trick. The drains were fixed.'

Next day, I went with Stella to visit some of the women who lived in caravans.

'I'm embarrased to let you in,' said the first woman. 'The van's in a bit of a mess.'

'Well,' I replied cheerfully, 'most of us don't live in places where we have to have the bedroom on show every time someone comes to the front door.'

'I feel like a second-class citizen at times,' said another woman we visited. 'The other day I went down to the library. They said that because I was not a householder, I would have to put down a heavy deposit before I could take out any books. When you tell people you live in a caravan, they look at you as if you had leprosy.'

It was nearly 11 a.m. and the sun was blazing down. Outside another caravan, a couple of children's bikes were leaning against the side wall.

'Where do the children ride?' I asked the mother, looking around the cramped space in the park.

'Not where, but when,' said the mother with a laugh. 'You can't let the children out when the sun's like this. They have to stay inside.' A look of resignation passed over her face. 'That's when I really feel the walls of the van beginning to close in on me.'

Another woman was even more forthright.

'What do you think it's like when your husband comes home in a bad temper, or,' she added with honesty, 'you are in a foul mood yourself. You can't go into the bedroom and bang the door behind you. You can't even go to the loo. It's fifty yards away and you have to stand in a queue.'

We all laughed at that, but I felt the frustration of someone who felt trapped in the confines of her caravan.

We left the caravan park and headed back to the town.

'I've heard how the mothers feel,' I said to Stella. 'But what about the kids?'

'You'd better ask the headmaster,' she answered, as she headed the car towards a school.

The headmaster expressed his concern about the caravan kids, as he called them.

'Some of them show signs of serious motor retardation. It's surprising how much walking the average person does in a house and how little someone who lives in a caravan does. The limited space and the intense heat which keeps children indoors means their physical activity is pretty restricted.

'Another thing I've noticed, is that some of them seem to have hearing problems. I think it has to do with the fact that you don't have to speak very loudly in a caravan.'

We left the school and Stella drove me back to Gray's house. He was interested in my reactions to life in a caravan.

'Not much privacy,' I commented.

'Ever had sex in a caravan?' he asked. I looked at him carefully and saw he was serious.

'Not that I can remember,' I muttered.

'Living in a caravan with three kids and your neighbours about two metres away, with thin dividing walls, creates some problems in that respect. You think about it.'

I did.

That night we ate a meal in the single men's quarters. The food was good. A man walked up to us. He was dressed in the standard uniform of the miners and his tray was heaped with food.

'You haven't worked hard enough to deserve a feed,' Gray said to him.

'Look who's talking,' the man threw back. He thumped his tray on the table and sat down opposite us. 'Anyway, when are you going to start this bloody university you're always talking about?'

I looked at Gray in surprise. 'What's all this about a university? It's kindergartens that people want here.'

'That's not entirely right,' replied Gray thoughtfully. 'Bruce here is very keen to do some more study.'

'That's because you talked me into it.' Bruce sounded belligerent, but I detected a note of acceptance in his voice.

'There are a lot like him,' said Gray, when Bruce had gone. 'They have done pretty well up here, but they have no formal qualifications. Bruce is really scared that if he tries to study, he'll fail. There are others, including a lot of women, who are just plain bored and looking for some intellectual stimulus.'

'So what's this about a university?'

'I'm negotiating a deal with one of the universities in Perth. What we want to do is offer a course which covers subjects of general interest. People will be able to finish it without the threat of pass or fail hanging over their head. But if they do well, the university is prepared to give them some credit towards a formal degree.'

'How will they run the course from 2000 kilometres away?'

'There's a number of professional people up here. Engineers, teachers and others. They say they wouldn't mind doing a bit of tutoring.'

'What about the students? They'll probably come from all over the Pilbara.'

'Well, someone will have to chase around the place, looking after them,' replied Gray, with a grin. 'I suppose that will be my job.'

'That's the trouble with isolation,' I said gloomily. 'To beat it, someone has to run. Anyway, there's one problem we have in the cities that you don't have to face up here in the bush.'

'What's that?'

'Migrants. Or as they say in the city, the multicultural society. It seems to me that the outback is the last bastion of the good old Aussie ocker.'

'Let's drive up to Port Hedland tomorrow,' said Gray. 'I think I have a surprise for you.'

As we drove the coast road to Port Hedland, Gray told me there were a few migrants in Karratha. 'I met a woman the other day from the Philippines. She is living in one of the smaller mining towns up here. She married a guy she met in Manila.'

I looked out at the empty desert country that rolled past with unrelieved monotony.

'He told her he lived in the country,' Gray continued. 'She said that was okay because she had been brought up in a rural area outside Manila.'

I had been to the Philippines and knew a little of the lush countryside near Manila. I looked again at the bare brown country of the Pilbara.

'She must have had a shock when she saw the country here,' I commented.

'She flew straight from Manila to Perth and then up here. It was a hell of a shock. No-one spoke her language, let alone understood her culture. She had never used an electric stove before and the food in the little supermarket was totally different. Her husband was a sympathetic sort of fellow and for a while she rang back home to the Philippines every night. I gather it was a case of more tears than words.'

We arrived at Port Hedland and went on to the satellite town of South Hedland. The big shopping centre was even busier than the first time I had visited. Gray took me to a shop which had the words, Migrant Resources Centre, stencilled on the window.

We went inside. A woman came to greet us.

'This is Rinke,' said Gray. 'She comes from Holland. But here, you can come from anywhere.'

'We've got over thirty-five ethnic groups in South Hedland,' said Rinke. 'It's a bit difficult keeping up with them all.'

'Which is the biggest group?' I asked.

'The Muslims. They come from a lot of different countries. Many of the men worked in mines before they came here.'

I had noticed a number of Muslim women walking around the shopping centre wearing their traditional dress. They were in distinct contrast to the other women, most of whom were wearing shorts or light dresses.

'The women must have some difficulties,' I commented.

'Yes, they do. As a matter of fact I was able to resolve a very difficult situation for them just recently.

'Muslim women are not allowed to uncover their heads in the presence of other men.'

Rinke paused and laughed. 'Would you believe it, the only hairdressing salon in Port Hedland is unisex?'

'How do you solve that problem?' I asked.

'I used to be a hairdresser once. I asked some of the women if they would like to come to my place for a hairdo. I was swamped.'

My colleague, Jon, who worked at Hedland, had come into the shop while Rinke was telling her story.

'Muslim people find our attitude to marriage very strange,' he told me. 'They are staggered at the number of divorces.'

'They're not the only ones,' I replied, remembering the amazement of some of my friends from mainland China.

'They think too many marriages break up without the couples actually trying to make a go of it,' continued Jon. 'One of them told me that Muslim philosophy is that in the long run, it's easier to repair than despair.'

We left the Migrant Centre shop and stepped back into the flow of human traffic through the shopping centre. I found myself looking at the people passing by with a new interest. Where had they come from? It occurred to me that none of them had been born here and all of them had travelled thousands of kilometres to come and live here.

I remarked on this to Gray, who was walking beside me. His reply, as usual, was succinct.

'We're all migrants here.'

That comment kept on coming back to me in the years ahead. There was a sense in which every person or family who went to live in one of the new mining towns in the outback was a migrant, irrespective of whether they had come from overseas or not.

One such town was Jabiru in Arnhem Land. In the public eye, Jabiru had been at the centre of controversy over the uranium mining issue. It was also in the hot seat in the matter of Aboriginal land rights. More recently, the region has been associated with environmental issues.

All of these have tended to conceal another, equally important, issue. How do people cope with living in a raw new remote region of the outback, where the weather joins distance and emptiness to test the fibre of people.

Jabiru is in Arnhem Land at the top end of Australia. Before the coming of the miners, Arnhem Land was almost cut off from the rest of the continent. The only inhabitants were Aborigines, who lived in a number of small communities scattered along the coastline.

The first mining operation in Arnhem Land was for bauxite and occurred on the Gove Peninsula, at the east end of the region. Jabiru was the second mining operation and it is located at the western end, several hundred kilometres from Darwin and a lot further from anywhere else.

'I won't be sorry to move out of this caravan.'

Eric Main had been living in a caravan for about twelve months, but so had everyone else who was engaged in the building of Jabiru. In Eric's case, he shared the van with his wife and three children. Eric was a padre who represented all the major Protestant denominations.

We had driven down from Darwin and were approaching the outskirts of the new development. We drove past a small building.

'That was the first store,' Eric told me. 'When we came to live here it sold Coke, ice-cream and a few canned goods. Milk came in twice a week. If you wanted meat, you had to order it four days in advance and even then you couldn't be sure if you would get it.'

We passed another small building.

'That's the school,' said Eric. 'When it opened, there were fourteen children.'

Further along was another steel-clad building. Fixed to the front was the well-known Australia Post sign. Outside was a solitary telephone box. Three or four people were waiting their turn to use it. They were not enjoying standing in the blazing sun. I commented on this to Eric.

'You ought to see it at night. Sometimes there is a queue of fifty waiting their turn. They all bring their eskies, have a drink and turn it into a social occasion.'

Later, I was to witness the same scene in many construction camps across the outback.

'How many people are living here?' I asked.

'There were about 2000 when I arrived. They were working on the development of the mine site and the building of the town. I think there were only four houses at that time. The others lived in caravans and the single men's quarters. We arrived just before Christmas. The wet season had begun and the work was beginning to wind down. It's impossible to do construction work in the wet.

'A week after we arrived, there was a real tragedy. A young fellow died suddenly of a heart attack. He had a wife and three children. It devastated them.' He paused and allowed the pain of the memory to pass. 'It devastated the whole community really. There were only a few families left by then. Everybody else had gone off for the Christmas break. Being near Christmas and with children around made it even harder. Then of course everyone was away from their own families and friends.'

'How do the people up here take to you being a padre?'

'In that case I was able to help the family and they were very grateful. There were a lot of other people who needed a bit of propping up too.' He laughed. 'Of course, not everyone gives me a good reception. There are some tough nuts here and some people who have had a bad experience with the church.

'There was one fellow who wouldn't have a bar of me. He just walked away whenever I tried to make conversation with him. Then the company asked me to run a course of business studies. I'm a trained accountant, you know.'

I didn't know.

'This fellow had enrolled for the course. When he first saw me come in to lead it, his face was a picture. But we're the best of mates now.'

We were driving past a row of long huts.

'That's the single men's quarters. They get a lot of young blokes up here. They pretend to be tough, but underneath, most of them are very lonely, desperately so at times.

'There was a fellow in that block,' Eric said, pointing to a row of huts. 'He had a complete nervous breakdown. Nobody had a clue that he was going through hell. After a couple of days, someone went looking for him. They found him lying on his bed, dead.'

The old Fitzroy Crossing, sometimes twelve metres under water, now replaced by a high level bridge.

Fitzroy Crossing's biggest turnout. The official opening of the new hospital in 1976.

Relic of the gold rush days. Railway station at Normanton, north Queensland.

We drove through the new town site which included an artificial lake. The site of the house Eric and his family were to occupy looked over it.

'Yes, I won't be sorry to leave the caravan,' he said as we picked our way carefully over the site. 'But it's been a great experience being in on the ground floor of a new community, something that the people who come after will never know.'

I commented on the style of the houses which had been completed.

'They're very comfortable, but they don't solve all the problems of living in this kind of situation.' He pointed to one of them. 'There's a woman living there. Her husband drinks heavily. She's an Aborigine and her relatives land in on her in large numbers. I think she feels that all the other women are watching to see if she'll cope. Some people are like that with the Aborigines, you know. They assume they will fail and never think to lend a helping hand.'

I left Jabiru feeling that urban transplants didn't solve all the problems of living in isolation and that people like Eric had a real role to play in that kind of community.

Apparently others thought the same way. The next time I heard of Eric, he had been appointed the first mayor of the new town of Jabiru.

Most of the mining towns of the outback still have the appearance of urban transplants. Some of them, like Tom Price, are no longer company towns. People who went there in the early days now own their houses, pay rates and grizzle like the rest of us. For them, Tom Price has become home.

The roads to the coast have been improved and it's not an uncommon sight to see boats parked in the front gardens of Tom Price, waiting to be taken for a weekend jaunt of 350 kilometres for a day's fishing in the Indian Ocean.

The children whose parents worried about them ten years before, went off to board in Perth and sometimes on to university. They come home to visit their parents as much as kids do anywhere.

'People who came up here to work in the early days soon came to know if they were going to cope,' said Sir Russel Madigan, one of the pioneers of the Pilbara development. 'They either got out quickly or stayed and settled.

'Those who stayed, came to enjoy the life. It can of course have its problems. If your mother lives in Melbourne and is eighty, falls and breaks her leg, it's a long way to keep going back. But people have stayed, survived and, yes, prospered.'

The words of Gray Birch about the new mining towns came back to me again. 'We're all migrants here.'

8

The Long Road Forward

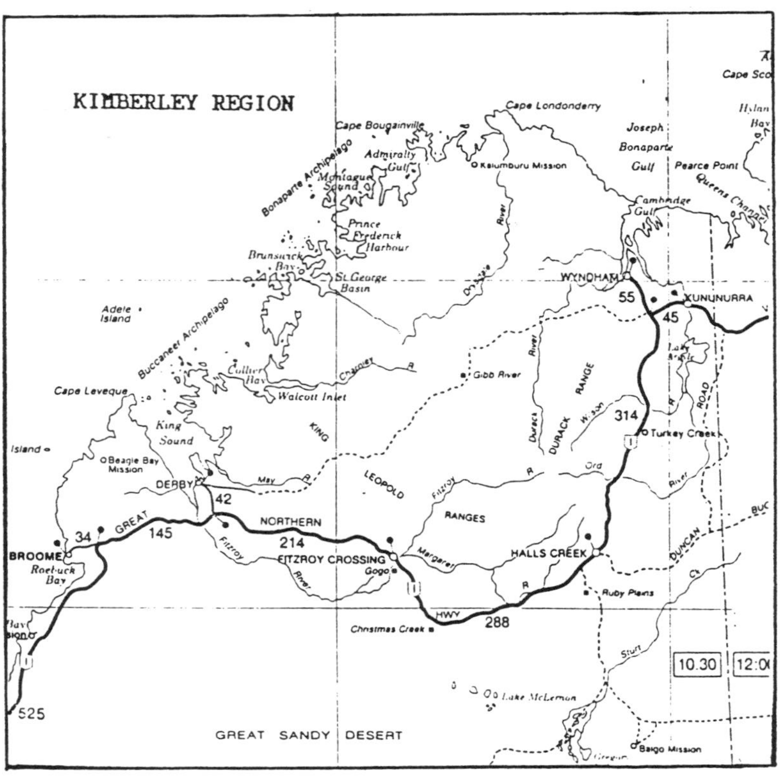

‘If there was a plane heading back to Perth now, I’d be on it.’

I looked up at the young woman who had spoken those desperate words. Her face was drawn and exhausted. She was nursing a very small baby and was obviously not long out of hospital. We were sitting in the waiting room of the Derby airport. It was fiercely hot and steamy. I pulled out a handkerchief and mopped my face.

'If there was another spare seat on the plane, I think I might join you,' I replied.

The waiting room at the Derby airport was a small weatherboard building. The floor was covered with worn brown linoleum and the walls were painted a hard white with bilious green trimmings. It was furnished with a few plastic chairs.

Apart from the woman and the baby, I was the sole occupant. We had flown up from Perth on a commercial flight. All the other passengers and the local airline staff had departed and switched off the air-conditioning.

'I hope you don't have to wait long,' I said. 'It must be hard on the baby.'

'My husband is flying in from the station to pick me up. He's probably been held up. He's terribly busy at this time of the year.'

'Why did you come back so soon?'

'Jim needs me,' was her bleak reply.

We heard a plane fly low overhead.

'That'll be Jim now.' The woman struggled to get up with the baby. I helped her take her bags out the door. The husband got out of the plane and walked over to us, greeted his wife awkwardly and cast a quick glance at his new child. Then he picked up the bags and walked quickly back to the plane. In a moment they were gone.

I walked back to the edge of the airstrip and sat under the shade of a tree. The sun beat down relentlessly.

Nine years before, I had come to the Kimberley in what someone had called, 'the good time of the year.' When I stepped out of the plane then, the sun had greeted me with warm invitation. I remembered vividly the breath-taking scenery, the broad sweeping plains and the strong spines of rugged mountain ranges, coloured in blue and purple and orange and green.

The visit I was now making was not in the good time of the year. The blazing sun was burning its mark on the landscape and on anyone who was foolish enough to venture out in it. The humidity kept everyone in a bath of perspiration and to shower and put on fresh clothes was an exercise in futility. I looked out across the empty airstrip and the dull brown mudflats around it and decided that the Kimberley region could not be taken for granted.

The gentle buzz of a distant aircraft pulled me out of my reverie. A busy little single-engine Cessna circled the airport as if defining its target and then dropped neatly onto the strip. I picked up my bag and walked across to it. Gordon didn't even bother to turn off the engine.

'Come on,' he shouted over the engine's roar. 'It's much cooler upstairs.'

I opened the cockpit door and jumped in. We took off. It was cooler upstairs.

Gordon was kept busy for a while, checking direction, mixture, trim and all those other mysterious rituals of flying. I had time to think about the place to which we were flying. My memories of the Kimberley countryside were vivid but there were other memories of Fitzroy Crossing which were as disturbing as they were vivid.

The woman with leprosy at Fitzroy Crossing hospital, the appalling conditions in which the Aboriginal people lived and the ominous prophecies of what would happen, if something was not done quickly to improve their lot.

I turned to Gordon, who had now satisfied himself that we were on the right track.

'Has it changed much at Fitzroy?'

Gordon was a padre with years of experience in the outback. He had been flying around the Kimberley for a long time. He looked at me and laughed.

'Last time I was down at Fitzroy the nurses lined up with a list of complaints as long as your arm. The diesel engine that generates the electricity had packed up. The other engine for the water bore had to be hand-cranked to start. You can imagine how they loved that.'

Gordon laughed again. I detected a note of cynicism.

'Oh yes. One of the outside drains had collapsed and the sewer was blocked. Those were just some of the things that had gone wrong on the outside.' He paused while he corrected the plane's heading. 'The inside was just as bad. Both refrigerators were out of action, including the one they use for storing drugs. There were a lot of minor things. It was enough to send the nurses screaming.'

I stared down at the wide open country below me. This was 1974. I wondered how many other nurses in Australia were doing battle with diesel engines to maintain their water and electricity supplies and had to put up with blocked sewers and drains and refrigerators which broke down in the fierce tropical heat.

Gordon interrupted my thoughts.

'Of course the real problem has been the upheaval in the community,' he said. 'Since the Aborigines received citizenship rights, they have moved away from the cattle stations in large numbers. They're living in camps around Fitzroy Crossing. It's created a heck of a lot of extra work for the nurses.'

'How's the leprosy?' I asked, remembering my first encounters eight years ago.

'Louise told me she was treating about forty Aborigines for leprosy at Fitzroy. And there are another four she had to send to the leprosarium at Derby.'

I looked down again on this awesome country which concealed so much human misery. Far below us was a tiny cluster of buildings which marked the homestead of one of the large cattle stations.

'We were down there a few weeks ago,' said Gordon, pointing to the homestead. 'They had an outbreak of scabies. It happened after the Aborigines came back from the races at Derby. Every camp in the district was affected. As fast as the nurses treated one batch of patients another came in.'

He paused again to check the plane's flight path.

'In the end, we had a meeting with the elders of the tribes. They got all the Aborigines together on the stations and in the camps around the town. Then I flew a team of nurses into the stations and another team did the town camps. The elders got the people to take showers and the nurses sprayed them with a delousing agent. We even burned all the bedding.'

'Did it work?'

'Yes it did. The nurses were exhausted, but the scabies have gone, thank goodness. Until the next races, I suppose.'

A line of trees snaked along the horizon in front of us.

'That's the Fitzroy River,' said Gordon. 'The airstrip's on that rising ground on this side.'

He began to manoeuvre the plane around the sky in preparation for landing.

'That's a grader!' I said, noticing a cumbersome machine moving slowly along the strip.

'Trust Reg to decide to work on it today,' grumbled Gordon. 'He's probably inside his cabin, with ear muffs on. How in the heck am I going to attract his attention?'

Gordon tried a couple of low level passes over the strip, but the grader driver remained blissfully unaware of our presence. Finally, muttering something about not hanging around up here for the rest of the day, Gordon brought the plane in at ground level, straight at the grader. We saw the driver's astonished face as the plane pulled up over him at the last minute. By the time we had banked and turned around, the grader was well off the strip and heading for the bush.

We landed and reluctantly stepped out into the blazing sunshine. Reg was waiting for us.

'Give us a lift into town?' said Gordon.

'After what you did to me? Come on.'

We climbed into the cabin of his truck and headed for the hospital.

Outwardly, Fitzroy Crossing had not changed. There were no more buildings in sight and the roads were as bad as ever. The hospital too, when we reached it, showed little sign of change. The old two-storeyed iron-clad building was as weatherbeaten as ever and I gathered from Gordon that the white ants still held dominion in the floorboards upstairs.

But immediately we went inside, the difference became apparent. There were nurses and patients everywhere. The nurses were obviously flat out and had no time for chatter.

'Both the hospital vehicles are out of order,' one of them told me. 'If you want to make yourself useful, you can try and get one of them in working order. We need one to get to the camps for the clinics.'

Not being a mechanical marvel, I went outside with some trepidation. I looked at the two old Land Rovers which stood in the yard of the hospital, and decided that the only solution was to cannibalise one of them to fix the other. It took time, but I was nearly finished and was about to put the wheels on when I discovered the wheel nuts were missing.

It was summer in the Kimberley and the temperature was having no trouble recording 45°C. A large placid-looking Aborigine had been watching me steadily lose my temper.

'Peter, where did you put the wheel nuts?'

'Dunno boss,' he replied cheerfully. 'Over there somewhere,' pointing vaguely in the direction of some long grass.

We trampled around in the grass for some time, with my temper soaring to match that of the atmosphere. Eventually we found them, and I went back into the hospital and reported to the nurses.

'One of the Land Rovers is working now,' I said tersely. 'But that Peter won't be working for much longer. The only little thing I asked him to do, he mucked up.'

The nurses looked at me in alarm and I suspected they were about to say something, but probably the look on my face caused them to leave it for a while. By evening, the temperature had cooled down and so had I. We were sitting around the kitchen table enjoying a meal which had been cooked by the nurses after a busy day in the clinic.

'You mustn't be too hard on Peter,' said Marcia, one of the nurses.

'Yes, well he's a nice enough fellow,' I replied. 'But the way you nurses work, you need all the help you can get and I reckon Peter is probably more harm than help.'

'He's more help than you know,' said Trudi, another nurse. 'If it weren't for Peter, we'd never get any sleep.'

'He sleeps at the hospital,' explained another nurse, Gail. 'The people here come around at all hours, no matter what the complaint. If we had to answer every call during the night we would be driven crazy.'

'Now they go and see Peter,' continued Marcia. 'If all they want is a headache powder, he sends them away. If it's serious, he comes and calls us.'

'So you can't sack him,' said Trudi with finality. 'We need him.'

Peter was one of a number of Aborigines who worked around the hospital whom I came to know. Another one was introduced to me as Horrible Wallaby, or Horrible for short. I never found out why. Even the other Aborigines called him that. Horrible watered the few patches of lawn and shrubs that had been planted around the hospital. He came and went in his own time and I don't think anyone really knew where he lived.

'Horrible's a maban, you know,' said one of the nurses, one day, as if that explained everything.

'What's a maban?'

'He's a kind of local medicine man. When we do a round of the patients, he comes with us. If we take a pulse, so does he, although I don't think he really knows what it is. His presence means a lot to the patients. I believe they slip off and see him even when we're treating them.'

I met some of the Aboriginal women in the hospital laundry. It was a simple building with a couple of very basic washing machines.

'We call them "cement mixers",' said Trudi, who was working with the women. 'The Aborigines haven't quite mastered them yet, but fortunately they are easy to fix.'

'Do you want to come out to Fossil with me?' Nancy asked me after we had finished lunch. 'I've got a clinic there this afternoon.'

Fossil Downs was a well-known cattle station, not far from Fitzroy Crossing and Nancy was one of the Community Health nurses. We headed down the road to Halls Creek and after a while turned off onto a side track. We bumped along through the bush for a while and then Nancy stopped the vehicle and jumped out.

'Come on,' she said. 'We take to the water now.'

In front of us, the track ended at the edge of a river. It was wide and flowing strongly.

'The river's up because of the rains,' explained Nancy. 'The causeway's under water. The boys from the station will be along shortly to take us across.'

We unloaded the equipment from the Land Rover and carried it down to the river. Then we sat and waited. After a while, there was the sound of a truck on the other side and it eventually appeared through the bush. Two young men hopped out and waved cheerily. Then they lifted a small dinghy from the back and placed it in the water. We watched as they rowed across.

'Gooday, Sister,' greeted one of the stockmen. 'Like a turn at rowing this afternoon?'

Nancy said she might conserve her energy for the children she would have to chase all over the camp. We loaded the equipment and rowed across to the other side, then loaded it onto the truck and drove to the homestead.

'They call this, "tailboard medicine",' explained Nancy, as she unpacked her equipment. 'I normally work off the back of my Land Rover when I'm doing a clinic at a camp.'

She took out some scales and hung them from a branch of a tree.

'What do I do with this?' I asked, picking up a large tin.

'Those are the vitamin biscuits. You can make sure all the children get one or two.'

By this time, several Aboriginal women had drifted down to where we had set up shop. They hesitated when they saw me, but Nancy quickly encouraged them to keep coming. She examined the children's ears and eyes for infection and weighed them, chatting all the time to the mothers.

I saw a very old Aboriginal woman sitting in the shade of a tree some distance away. I went over to talk with her.

'She's over a hundred,' called Nancy. 'But a sharp mind.'

'I'm all right,' replied the old lady when I enquired after her health. 'But I'm having trouble with my eyes.'

'Show me your glasses,' I requested.

She handed them over. They were thick with mud. I took them across to a water tank and washed and polished them vigorously.

'Try them now,' I suggested.

The old lady put them back on. A look of sheer pleasure came over her face.

'Glory be!' she exclaimed. 'It's a miracle. I can see!'

I remembered the words of one of the pioneers of community health: 'When you clean up a person's eyes or ears, you not only restore sight and sound. You open up the beauty of the world. That's as important to health as any medical procedure I know.'

Nancy finished her clinic. The Aborigines who had been gathered round talking with her and with each other drifted away. It was a little more relaxed than the outpatients' department of a big city hospital. We packed up and called into the homestead and had a drink with the manager and his wife.

Later in the afternoon, the stockmen took us back to the river and rowed us across. We repacked the equipment into Nancy's vehicle and headed down the track to the main highway.

'Do you enjoy this better than hospital nursing?' I asked her as we drove along.

'I reached the stage where I was just fed up with patching up people, the same people, all the time,' she replied slowly. 'In Community Health nursing, at least there is the feeling that you are tackling the roots of the problems. Still, even now I have to do a lot of treatment at the camps. I have achieved some good preventative programmes, especially for the children. But until Aborigines adopt sound health practices we will still only be carrying on band-aid medicine.'

The trip to Fossil Downs was one of the shorter ones Nancy and her colleagues made every week. Some of them involved travelling hundreds of kilometres over rough tracks and staying out in the bush for two or three nights at a time.

'You must be glad to collapse into bed after the long trips,' I remarked.

'Blissful thought! Last week, I did one of the long trips, Christmas Creek, Cherubin, out that way. I got a flat tyre and had to change it by myself. Pulling the spare down from the roof of the Toyota is no joke. When I got back, the vehicle had to be cleaned. It gets absolutely choked with dust. Then there was a mountain of washing.

'And when that is all done,' she said, with a touch of weariness, 'there's the book work. That's the most tedious of all. We have to keep very detailed files and that isn't easy. You ought to try filling out forms on the bonnet of your car and brushing away a million flies at the end of a hot tiring day.'

'Anyone for a drink at the pub?' I asked.

It was after dinner and I remembered from my previous visit that the pub was where people went to meet people. It was the only place at Fitzroy Crossing where the nurses could go to get away from the hospital for a while.

To my surprise, there were no acceptances.

'What's the matter?' I asked in mock dismay. 'Can't you bear to leave the hospital?'

One of the nurses stood up.

'You had better come and see for yourself,' she said.

We drove down to the pub along the bumpy track, pushing our way through dense dust which had been churned up by other vehicles. As we pulled up, I noticed a crowd of Aborigines milling about outside in the dark. Before we had a chance to get out, an Aboriginal woman came running over.

'Sister, Sister,' she exclaimed. 'Been a big fight. Two fellers hurt.'

With a sigh of resignation, the nurse walked over to the crowd. I followed her. One Aborigine was lying on the ground. Another was bleeding profusely from the face. The nurse bent down to inspect the man on the ground. Then she straightened up.

'Will you help me get these two into the vehicle?'

Together, we managed to get the two injured men into the back of the Land Rover.

'Can you see why we don't go down any more?' the nurse said as we drove back to the hospital. 'This happened every time we went. Or if not we sat there wondering when a fight would start.'

Next morning, I was introduced to a young Aboriginal woman, Polly, who worked with the Community Health nurses.

'What kind of work do you do?' I asked.

'I go with Sister Margot to the camps. I fix people with sores and give old people their vitamin tablets and protein biscuits. I talk to the mothers about their babies and I make sure the leprosy people are having their treatment.'

'Are you the only one doing this?'

'No. There's January. She drives her own Toyota,' Polly added triumphantly. 'I'm learning to drive too.'

Later I met other Aborigines who were getting involved in the health process. There were Dicky and Friday, who were driving the nurses out to the stations. Well, you don't make neurosurgeons overnight and it was a long way from the steering wheel of a Toyota to the operating table in a major hospital. But it was a beginning and more importantly, it was a beginning in the right place.

'I hope the next town hasn't been turned upside down like Fitzroy,' I said to Gordon.

We were flying to Halls Creek, 200 kilometres further inland from Fitzroy Crossing. I remembered it as a pleasant little town, laid-out streets, a few shops and a better hospital.

'It's still good,' Gordon replied, 'although there are many more Aborigines around and the health problems are just as serious.'

We were going to Halls Creek to open a new kindergarten, which was probably unique in Australia. It had been built especially for Aboriginal children and proved to be very successful. On the official opening day, the whole town turned out. I watched Aboriginal mothers playing with jigsaw puzzles and looking with fascination at the child-size kitchen.

'You can blame the nurses for this,' said Gordon. 'They reckoned that unless we gave Aboriginal children a good start in education, they would never catch up.'

That night the nurses gave a party to celebrate the opening and to farewell one of their colleagues. In the bush, any excuse for a party is a good excuse. It was a barbecue, with the meat brought in from one of the nearby stations. Many station people had come in, because the nurse who was leaving had been popular. There were a lot of people from the town too.

'Have you met Brem?'

I looked up and saw one of the nurses dragging a reluctant man by the arm to meet me.

'No I haven't, but I've heard about him,' I replied, extending my hand.

Everyone in the Kimberley knew Brem. He was a dentist who lived in Wyndham on the coast, but travelled all over the region. He was a much loved character and stood and chatted for a while. Then I noticed a woman from one of the surrounding cattle stations bearing down on us.

'I've been looking for you for eight years,' she told Brem in a loud voice. 'You ruined my underclothes and you're going to pay for them!'

The conversation all around us took a quick moratorium. Nobody was going to miss this.

'Eight years ago, you were down here pulling out someone's teeth,' she went on. 'There was a big storm brewing up and people were telling you not to drive back to Wyndham, but you set off anyhow.'

By this time, everyone was riveted to attention.

'They got a call from Wyndham the next morning to say you hadn't arrived. All hell broke loose down here. They thought you must have drowned. Someone contacted our station and asked if we could start a search for you.' By now, the woman was in full dramatic flight. 'I saddled my horse and rode towards the Wyndham road. I had to cross two creeks to get there. We looked everywhere and couldn't find you, so I crossed the creeks again and went back to the homestead. By this time I was saturated. Then we got a message to say you were safe.'

She paused. Brem by this time, was sitting with his head down, looking for the proverbial hole in the ground to crawl into.

'And where had you been all the time, you wretch?'

'Playing poker,' muttered Brem.

Eight years later he met his nemesis in the form of an outraged woman whose underclothes had been ruined crossing the creeks while searching for him. Without a word, Brem pulled out his wallet, extracted a note and solemnly handed it to her.

'They've been saying that for years,' barked Cameron. 'Someone comes up from Perth. They fly into Fitzroy, look at the hospital, shake their heads in concern, tell the nurses what a wonderful job they are doing, make all sorts of promises, fly out the same day and that's the last we hear of them.'

We were sitting on the wide verandah of Cameron's homestead in the late afternoon. It was several months since my last visit to the Kimberley. In the intervening period I had extracted a promise from the Western Australian government to build a new hospital at Fitzroy Crossing. I had come out to this station, which was not far from Fitzroy, to tell Cameron and Liz the good news. Cameron obviously had heard it all before.

'I think we might have it this time,' I said. 'We've approved the designs. They go to contract next month.'

'Where will it be located?' asked Liz.

'Well there are those who want to retain the present site, and there are those who want to move it to higher ground.'

'Yes, well when the river gets up it's a hell of a job getting to the hospital,' growled Cam. 'Last year we had to take in a sick Aborigine. It was flooded all the way to the hospital.' He allowed himself a half grin as he remembered the occasion. 'It was all right going there. The current was behind us and very strong. On the way back we were almost here when the damned motor cut out. The current took us all the way back to the hospital.'

'I've asked for a boat for the hospital.' I commented. 'They think we are mad asking for it. They must have some idea of us being in the middle of a big desert.'

'There'll be big trouble if you try to move the hospital to another place,' warned the nurses, when I returned to the hospital. 'People don't want it moved.'

There was big trouble. Public meetings were held. There were two versions of what the Aborigines wanted. In the end, it was decided to move the hospital to higher ground.

'Now we can get on with the building,' I said with relief. 'The contract price seems too good to be true.'

It was!

'I'm afraid I have bad news.' It was a government official from Perth. 'The builder who submitted the successful contract had gone bankrupt.'

'What will you do?' I asked in dismay.

'We'll probably take over the work ourselves and subcontract out.'

I groaned inwardly. It was a recipe for disaster.

The months that followed were agonising, especially for the nurses at Fitzroy. With a new hospital imminent, nobody was keen to spend money on the old one, but progress was tortuously slow. Subcontractors came into Fitzroy Crossing, did a bit of work and then disappeared. Government departments were not geared for cracking the whip. The project was an unwanted nuisance. Fitzroy Crossing was nobody's baby. As time went by, the cost of completing the building escalated. It was a financial embarrassment to the government.

'You had better start preparing for the move.'

The man from Perth spoke the words we had been waiting months to hear. It was near the end of the year and the beginning of the wet season. After all the stress and disappointment of frustrating delays, tempers were frayed and the nurses were tired. In those circumstances, the coming of the wet season was the final straw. I flew up to Fitzroy Crossing to lend a hand. The nurses were in a state of exasperation.

'Do you know what it's like moving a hospital?' demanded Elspeth, an experienced bush nurse. 'We have to keep things unpacked to keep the hospital going. After all, the next nearest is in Derby.'

'When will the move happen?' asked Margaret. She, too, had been in the bush a long time.

'I wish I knew.'

My response was an expression of my own growing exasperation, but it was nothing compared with the nurses'.

'It's nearly Christmas, and if it doesn't happen in the next week it won't until the end of January.'

Christmas came and went. The waiting became intolerable. Breakdowns were more frequent. Everyone seemed to choose this time to become ill or have an accident.

'This one will have to go to Derby,' said Elspeth, one night, after examining a patient who had been brought in. 'We'll have to go by road.'

A storm had broken over the East Kimberley and flying was impossible. I didn't think that driving to Derby would be much of a jaunt either. One of the men around the hospital offered to drive with the nurse. They set off in the teeming rain to drive the 250 kilometres to Derby.

'There's an ambulance coming out to meet you half way!' shouted a nurse, running out of the hospital, just before the vehicle headed off.

It was very late at night before the nurse and the driver returned.

'The rain was so heavy, we missed them,' explained the driver. 'When we realised there was something wrong, we waited for a while. They came tearing back and nearly missed us.'

'This time we hope all will be well.'

It was the man from Perth again, telling me to prepare the hospital for moving. It was the end of January.

'I have four very frustrated nurses up at Fitzroy,' I said sternly. 'It had better be right this time.'

The nurses packed again and waited.

'We seem to have lost the electrical subcontractor.'

The voice of the bureaucrat from Perth had the same tone as if he had suddenly mislaid a paperclip. So I went up to Fitzroy Crossing again. The nurse needed someone to kick and I was the only one available.

It happened six times before the move was finally made. By then the nurses were about at the end of their tether. But the new buildings, air-conditioned, well-equipped for clinical work and with separate quarters for the nurses, were worth the wait.

There had been some concern as to whether Aborigines would come to such a sophisticated building after the simplicity of the old one. But they came and adjusted without any trouble. It was the station people who found the change difficult to accept. The new hospital was not a nice place where they could blow in, throw a sack of meat on the kitchen table and sit down for a cup of tea and a yarn.

The official opening of the Fitzroy Crossing hospital was the beginning of a new era. It was fifty years since John Flynn had recognised the need for a hospital at Fitzroy Crossing. The one he built in 1939 had been a simple medical outpost, manned for nearly forty years by a succession of dedicated nurses.

Now a modern hospital, with a separate Community Health facility and proper nurses' quarters, was to be opened. It was probably the biggest event in the history of the district.

Not far from the hospital, we had built a preschool centre, which had been designed particularly for Aboriginal children. It was due to be opened at the same time.

The third building to be opened in Fitzroy Crossing at this time was a power station. The hospital, the kindergarten and the power station were to be the first buildings of the new town of Fitzroy Crossing.

At the same time we had been building a new manse in Kununurra, for the flying padre. Gordon and his family had lived for years above the church. It was not air-conditioned and when the temperature was running at 45°C for weeks and the humidity was about 100 per cent, life in the upstairs flat was not very pleasant.

The Premier of Western Australia, Sir Charles Court, was coming up for the weekend of official openings, as were several federal and state ministers and government officials. My major concern was to get all the visitors from Kununurra to Fitzroy Crossing on the Sunday morning.

It was at this point that Gordon enunciated his 'plan for disaster in the outback, because that's what will happen' philosophy. He rang me in Sydney and said, 'You'd better bring another plane across.'

'Why? You've got enough planes in the Kimberley to move an army.'

Gordon insisted, so I reluctantly agreed. We flew in an additional twin-engine plane from Sydney.

The opening of the manse went well. After the ceremony and the celebrations, Sir Charles Court and his party flew to Wyndham to attend to some business. They had a twin-engine plane and would fly straight to Fitzroy Crossing the next morning for the other three openings. The rest of us went to bed on Saturday night, contented with the day's work.

I had hardly opened my eyes on Sunday morning when there was a banging on the door. It was a government official with a message from Wyndham. Could we please find Sir Charles Court a plane? The pilot at Wyndham had gone out to his plane early in the morning and found the cabin door would not close. You can't fly a state Premier around the Kimberley with the door hanging open.

The people in Wyndham had rung everywhere. Port Hedland, even Darwin. No plane was available. Eventually, Sir Charles Court exploded.

'Try the AIM. They'll get me one.'

'Isn't it lucky we brought an extra plane?' said Gordon, smugly.

We flew across to Wyndham and gave the grateful Premier our plane. Then our pilot had the bright idea that if we screwed up the cabin door of the other plane we could crawl in through the little luggage hatch and fly down to Fitzroy Crossing. The rear view of one or two of the passengers squeezing in through the luggage compartment provided some light relief.

In due course everyone arrived at Fitzroy Crossing. The power station was officially opened in the morning. The hospital opening was held up by the late arrival of the

Federal Minister for Aboriginal Affairs. We had over 500 people, most of them Aborigines, standing in the hot sun. It was time for another bright idea.

Sir Charles Court, who was not given to sitting around, said to me, 'We'll reverse the proceedings. We'll have the inspection first and by then the minister will surely have arrived.'

So 500 people poured in to inspect the hospital, which was the real reason they had come. Then the minister's plane landed and we got on with the official opening.

Finally we all trooped down to the kindergarten. Johnny Marr, one of the Aboriginal elders, gave the address and formally opened the kindergarten.

Eventually the captains and the kings departed and we were left to ourselves. We had decided to have a barbecue at the kindergarten with just the workers indulging in their own private celebration. Everyone went to shower and clean up. In the cool of the Kimberley evening we strolled down the hill to where the smoke was softly rising from the barbecue. Those are the times in the bush when it is at its best.

Then to our amazement we saw a large articulated truck grinding up the road to the hospital. As it was Sunday evening, we had no idea what it could be. So we went across to find out. The driver got out.

'Is this the hospital?' he asked. We said it was.

'Well I've got your morgue freezer on board. It's a bit big. Can you give us a hand?'

We helped the driver get it down and duly installed in the morgue. The driver didn't take much encouragement to come and have a drink and a steak.

A little later, in the darkening of the Kimberley evening, we witnessed another piece of history in the making. The street lights of the new town of Fitzroy Crossing were turned on from the new power station. We all stood silent for a while as we tried to absorb this new wonder. The lights twinkled bravely in the blackness of the night.

Suddenly from the dark distance, a voice came across the still air.

'She'll never be the same place again.'

We stood silent as we took in the implications of that comment. I thought about all the nurses who had wrestled with the erratic diesel engines of the power plant at the old hospital. For better or for worse, Fitzroy Crossing would never be the same again.

9

Distance, Devastation and Darwin

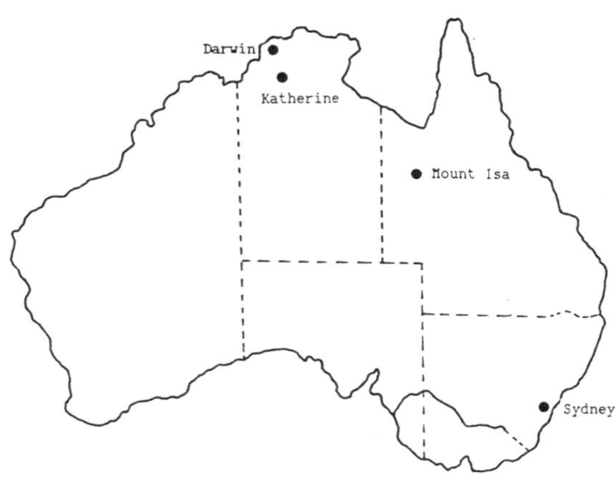

Darwin was always different. They called it the town with the beer can mentality. Even its major sporting event was a regatta with rafts made of beer cans. History has it that there were three attempts to find the right location for Darwin and the cynics added that even the third time wasn't lucky.

Darwin was always different, but it took a tempest they called Tracy to demonstrate that the real difference was Darwin's distance and isolation from the rest of Australia.

Christmas Eve 1974. Offices, factories and shops all over Australia were beginning a last-minute frantic clean-up before closing for the festive season. People were emptying into the streets and scurrying for buses, trains and trams to take them home. Last-minute calls were made to business associates to wish them a Happy Christmas.

I was making one of them. My call was to a colleague in Darwin. Bernie and I had seen a lot of each other during 1974, working mainly on Aboriginal projects in the north. I rang to thank him for his cooperation.

'How's the weather?' I asked casually, guessing what the answer would be. The temperature would be its usual monotonous 33°C and the atmosphere would be saturated with oppressive humidity.

'Pretty much the same as usual. There's supposed to be a cyclone hanging about off the coast,' he added. 'But I expect it will go away shortly. Just stay long enough to give us something else to think about.'

We wished each other a Happy Christmas and rang off. The importance of Bernie's casual reflections would rebound savagely within the next twenty-four hours, but we didn't know it at the time.

Before packing up for the day I sat back for a moment and tried to visualise Darwin and its unique lifestyle. Tomorrow, many of its people would be celebrating out of doors, sitting back on folding chairs in shorts and thongs. The men would be clutching a stubby, and the women a glass of beer or white wine. Friends would be dropping in, casually dressed, casually greeted. 'Help yourself to a beer. Grab a plate and pull up a chair.' Overhead, the sky would be glowering grey, with the threat to spoil all the fun by turning on a thunder storm. But the people of Darwin had seen it all before.

I pulled down the roller top of the desk that had been John Flynn's, closed the office door and went home.

Christmas Day and Christmas dinner. The family had been rounded up and arranged in their respective places at the table. The conversation began to flow freely.

The phone rang. The family sighed. I got up to answer it.

'It's Harvey,' said the voice at the other end. 'Have you heard the news?'

'What news?'

'About Darwin. It's been hit by a cyclone. There's not much information through yet, but it sounds as if the city has been wiped out.'

It was the sort of information we were accustomed to hear about places overseas. I remained silent as I tried to grapple with the thought of an Australian city totally destroyed.

'What about our people?' I asked, remembering Bernie and his casual reference to the cyclone. Darwin was the headquarters of our work in the north. I had a lot of friends who worked there.

'We have no idea. That's why I'm ringing. We think you should try to get up there as quickly as possible. Do you think you can get there?'

'I'll have a go,' I muttered, not having the slightest idea how I could get there. Obviously, all hell was breaking loose over Australia. I went back to the table and told the family. Someone turned on the television. It told of Cyclone Tracy and confirmed the seriousness of the situation, but had little detail as to what had happened.

One item on the news captured my attention. 'The RAAF is gearing up for a massive operation. It may be necessary to evacuate the whole population of Darwin.'

I thought for a moment. There would be a lot of planes flying to Darwin to bring people out. Maybe there would be a space for someone who wanted to get in. I rang the RAAF base at Richmond and explained why I wanted to get to Darwin.

'We're sending planes up from early tomorrow morning,' I was told. 'Come out as early as you can. I can't promise anything, but we may be able to slip you on board a Hercules.'

I drove out to Richmond at 4 a.m. on Boxing Day. The darkness concealed the fever of activity that was happening. The big bulky shapes of the Hercules transport planes stood impassively on the tarmac as the ground crews fussed around them.

'Just keep out of the way,' ordered the Commander, when I caught up with him. 'We'll sneak you aboard at the last minute.'

I made myself inconspicuous and waited apprehensively. The darkness before dawn was just beginning to lift as I climbed through the tail door into the cavernous belly of the first plane to fly out from Richmond to the beleaguered city of Darwin. The tail door slammed shut and the silent emptiness inside the plane was shattered by the opening roar of the Hercules' four engines. We lifted off and began the long journey north.

I released my seat belt and turned my attention to what was happening around me. It was an interesting and intriguing sight. Pride of place was occupied by a truck with a canvass canopy. Inscribed on the side of the vehicle, were the words, 'N.S.W. Police Rescue Squad'. It was attended by two policemen who looked as if they would be capable of rescuing Darwin by themselves.

'What kind of equipment are you carrying?' I asked innocently.

They looked at me as if I had asked for the key to a bank.

'That truck,' said one of them, 'is loaded with guns and ammunition. There's a fear that looting might break out.'

I moved further down the plane to where a couple of navy officers were sitting.

'The aircraft carrier *Melbourne* has been ordered to steam to Darwin to help with the relief effort,' explained one of them. 'We're going up as the advance party.'

The other officer laughed. 'Actually,' he confessed, 'we were at a party in Canberra and missed the boat.'

There was a quiet woman sitting by herself.

'I'm a nurse,' she explained. 'I was on leave from Darwin Hospital. I heard the news and felt I should return.'

The other passenger to whom I spoke introduced himself as a meteorologist. I resisted the temptation to suggest that his going to Darwin was shutting the door after the horse had bolted.

Some of us had been to Darwin before. We talked anxiously of the people we knew and our concern for their safety.

'It's those houses on stilts that worry me,' said one of the navy officers. 'They are very vulnerable.'

We landed at Mount Isa and refuelled, because there was no way of telling whether there would be fuel available at Darwin. The flight took us over Katherine and soon we were about 100 kilometres from Darwin.

'Come up on the flight deck,' invited one of the crew. 'You'll see better from there.'

We stood on the flight deck and strained our eyes towards the horizon, waiting for the first glimpse of the city. When it came, it took us by surprise.

'What are those silver things wrapped around the trees?' asked the nurse.

'Sheets of corrugated iron,' observed the pilot. 'They must have been ripped off the roofs.'

I looked down at the tiny toy-like trees and the iron that was like tinsel paper wrapped around them. From this height it looked more like a Christmas party than the aftermath of a cyclone.

'Look at the houses,' breathed the co-pilot.

Whether the houses looked toy-like from this height or not, destruction is destruction in any dimension. We were looking at massive destruction. As we flew across Darwin

Devastation in Darwin. Stilt houses destroyed by Cyclone Tracy, Christmas Day 1974.

Kindy in the bush. Pat Teasdale, pioneer mobile preschool teacher in the West Kimberley region, with station children.

towards the airport, it dawned upon me that I was looking down on a destroyed city. The city of Darwin was dead.

Then I saw a car, also toy-like, moving slowly down one of the streets. It was the first sign of life.

The Hercules banked steeply and came in to land. As we taxied towards the terminal, I looked across to the tarmac where the light aircraft usually parked. The scene was unbelievable. There were planes lying on their sides, like birds with broken wings. Others were upside down. One had been caught up by the blast of the wind and finished on top of a hangar. It was as if an angry God had snatched up a handful of planes and thrown them across the landscape.

The main terminal building at Darwin airport had always been an eyesore and the cyclone had done nothing to improve it. The Hercules wheeled away from it to the right and came to a halt outside the RAAF base. The tailgate came down and we stepped out into the eerie atmosphere of the aftermath of a cyclone. Heat rose from the tarmac beneath our feet. Desolation hung in the air. The air force people looked as if they had taken up action stations in a war.

'Anyone going into town?' I asked.

'I'm just going in now,' said one of the officers. 'I'll give you a lift.'

We headed towards the centre of the city. The grey heavy skies overhead added to the eeriness of the scene.

'Been here before?' asked the officer, as we drove along.

Without taking my eyes from the scene I said, 'Several times'.

'It's like this everywhere,' he continued, looking at the shattered houses that lined the streets. 'The stilt houses copped the worst of it. Some of them were swept away like a house of cards.'

Every building we passed bore testimony to the indiscriminate ruthlessness of the cyclone. The lush tropical trees and bushes which were a feature of Darwin had been smashed to the ground.

We came towards the centre of the city. I had noted from the air that Darwin's tallest buildings seemed to be intact. I asked the driver if they had been damaged.

'The glass has gone. But they look as if they survived all right. We won't know for a time whether they've been structurally weakened.' Suddenly he laughed and pointed in the direction of a multi-storeyed hotel. 'Look over there.' A car had finished up in the swimming pool. 'That'll take some explaining.'

We came to my destination. I thanked the officer for his help and got out. In front of me was a large church. It had been built with an A frame and the girders had been constructed of iron. Apart from some smashed glass panels, it seemed to be unharmed.

Not so, some of the buildings beside it. One of them had taken a savage beating and the grounds around it were strewn with wreckage. It was one of ours and, although old, had been like a good and faithful servant. To see it so wantonly damaged was upsetting.

Then I turned and walked towards the building next door. It was a low tropical bungalow house, surrounded by a garden. The last time I had seen it, there had been a row of tall banana trees running down one side. They had been ripped apart and were now a heap of torn branches lying on the ground. The once-neat garden was filled with rubbish and wreckage. Part of the roof had been torn off, but the walls seemed to be standing.

I picked my way through the wreckage and stepped onto the verandah. The front door was open but there was no sight nor sound of life. I stood for a moment, fearful of having to make the first step towards discovering what had happened to my friends. Finally, I banged on the door and called out, 'Doug, are you there?'

The sound of my voice echoed back to me and then sank back into the silence. Then I heard the slip-slapping of someone walking in thongs. A figure began to emerge from the gloom and a cheery voice rang out.

'Max, how nice of you to come.'

Doug was obviously alive and well. It was a promising start. We greeted each other and I explained how and why I had come to Darwin.

'We thought you might,' said Doug.

'And where were you when the lights went out?' I asked.

'Actually I was in the bath. Not I might add, having a bath. But when the wind began to tear apart the roof and the walls, we thought we had better find somewhere for protection. The bath seemed to be the best place, but we pulled a mattress over the top of us just in case. It was lucky we did.' He pointed to a large javelin-shaped piece of wood that lay on the floor. 'That brute apparently came through the wall.'

I looked at the piece of timber. It was about three metres long and sharply pointed at one end. I told Doug I was glad he had not been skewered.

'What about my other mates?' I asked. 'Are any of them hurt?'

'Brian lost his wife and little daughter,' he replied slowly.

'What happened?'

'They went down into the concrete room beneath their stilt house. Most of the people did at the height of the cyclone. Brian was sitting on the floor with his back against the wall. His wife and daughter were sitting on the other side of the room.' Doug paused and then went on. 'The wall behind them just collapsed. Brian sat and watched it happen. He couldn't do a thing. They were killed in front of his eyes.'

We simply sat for a moment in silence.

'How is he?' I asked eventually.

'He's in a state of shock. We're just leaving him be for a few days and hope he will come out of it.'

There was a sound of footsteps on the verandah and the shadow of a visitor blocked the light from the doorway. I looked up and saw Graeme, another of my Darwin mates.

'Thanks for coming,' he said, as casually as if I had just dropped in from over the road. It was how I would feel for the next few days; a neighbour coming in to express sympathy on the occasion of a death in the family.

'How's your house?' I enquired.

'You'd better come and see,' said Graeme in his customarily serious voice. 'Doreen is working flat out at the hospital. They need every nurse they can get.'

As we drove to one of the outer suburbs where Graeme and Doreen lived, we passed streets where every stilt house had been wiped out, right down to the floorboards.

'They're already calling them "billiard table" houses,' remarked Graeme drily. Then he pulled up the car.

'There it is,' he said. 'Or rather, there it was.'

There was nothing to see except the bizarre sight of the stilt foundations and the bare floor. It was no different from the other houses, similarly destroyed. But I knew this house. I knew the way in which Graeme and Doreen had furnished it, the possessions they had gathered over the years. It was all gone.

I turned to my stoical mate. The tears were running down his face. I suddenly realised I was crying too.

Feeling a bit embarrassed, I turned and looked at the whole monumental devastation. Street after street of shattered houses, trees torn down to stumps, light poles tilted drunkenly. An occasional car came cautiously down the street, the drivers stopping from time to time, looking at the wreckage and shaking their heads in disbelief. It was hard to absorb the whole of the horrible reality.

We got back in the car and drove back to Doug's place.

'So. Where do we start?'

'There's no water, electricity or sewerage,' replied Doug. 'We'd better get some water. They're tapping straight into the pipeline.'

People who have been to Darwin will remember that one of its less attractive features is the rusty old water pipeline that ran above the ground alongside the main highway. After Cyclone Tracy, it proved to be a godsend. Some enterprising spirits drilled holes into the pipe and screwed in taps. People drove up in their cars towing trailers carrying forty-four-gallon drums and anything else they could lay their hands on, to fill with water.

We waited in the queue and filled out containers. Then we went out to see how some of our other friends were faring. We found Gordon in the front garden, stripped to the waist, digging a trench.

'It's for a temporary latrine,' he explained. 'We don't know how long the sewerage will be out of action.'

'What's the general situation?' I asked.

'You heard about Brian's tragedy?' he asked. I nodded.

'The rest of our staff and their families are okay. Although they're all still pretty shaken. We've got everyone working, because that's the best way. But every now and then one or two of them just go away and have a quiet howl.'

'What about the property?'

'We've got about twenty buildings around Darwin. They've all been damaged. You've probably seen some of the worst.'

We left Gordon to return to his digging.

'You heard about the hostel?' said Doug. 'We built it for Aboriginal people and named it after Gordon. He's been up here for twenty-two years you know. We opened it last Saturday. Gordon was as proud as anything. The cyclone wrecked it.'

For the first time I heard a note of bitterness in his voice.

We drove on and stopped outside the home of my mate Bernie, who had made the casual reference about the cyclone on the phone on Christmas Eve. The damage to his house was less obvious, but the shock to the residents was apparent.

Close by, we heard a squeal of tyres, followed by the sound of gun fire. Then there was silence. We learned later that this was one of a few incidents where the police had apprehended looters.

We drove past the remains of houses where people had already set up makeshift camps and were beginning to sort through the wreckage in the hope of salvaging something.

'They're afraid that it will rain,' commented Doug, 'and ruin all the things they can't put under cover.'

But the people we saw weren't panicking. They knew that the extreme heat and humidity of Darwin quickly put an end to hustle and bustle tactics. So some of them were just sitting around having a drink. They waved to us as we passed. I had the feeling that they were just beginning to realise they were alive and that was worth savouring for a moment.

In the gathering dark, we drove back to Doug's house.

'I suppose we'd better think about getting a meal,' he said.

'Well, I wasn't sure what to bring up from Sydney,' I replied apologetically, 'but some of my neighbours came in with some biscuits and tins of food.'

'Oh, I think we can do better than that,' Doug said airily. 'We had the freezer filled for Christmas. It'll be good for a day or two yet.' He turned to me and grinned. 'How about some nice barramundi or buffalo steak?'

I built a fire and before long we were settling back to relax and enjoy some good food. Since there was no electricity and we had to conserve the torches, we turned in for the night.

It had been a long day since I had driven out to Richmond Air Base in the early hours of the morning.

Next day we continued with our inspection of the shattered city. There were some bizarre sights. Fanny Bay gaol was one of them. On top of the high surrounding wall was perched a toilet bowl. The guardhouse, of which it had been a part, was blown away.

'I always wondered what the guards did,' said Doug as we drove past.

We turned down a side street. In the front garden of a battered house stood a boat. Tacked to its hull was a sign which read, 'Tracy couldn't remove me and neither can you.'

Further along we stopped to talk with some people whom Doug knew.

'How did you get on?' I asked the man, who had been rummaging around in a pile of wreckage. He wiped his hands and twisted his head to look at the house behind him.

'Well, you can see what happened,' he said. The hurt of it narrowed his eyes and tightened his mouth. 'But we escaped without injury. The neighbours weren't so lucky.'

I asked what had happened.

'They were upstairs at the time,' he began, looking up at the empty space where the house had been. 'There was Bill and Jan and the two kids. The windows went first and then the walls started to go. Bill said they'd be better on the ground, so he took the kids, one at a time, and crawled down the outside steps with them underneath him for protection. He put them on the side of the barbie out of the wind, for protection.'

He looked across to the brick barbecue in the next yard, as if he still couldn't believe it.

'While he was taking the second kid down the wind ripped into the walls and they started to blow away. Jan grabbed a door handle. Clutching at a straw, I suppose you'd call it. But the door suddenly came off its hinges and took off with Jan still hanging on for dear life.'

He gazed into the air as if seeing it all happen.

'The door finished up in the yard on the other side. God knows how Jan survived, but she did. She lay there all night with a broken ankle, with Bill and the kids only a few yards away. They had no idea what had happened to her until the next morning when the wind had calmed down.'

We left the man to continue with his clearing.

'I'll leave you with Bernie for a while,' said Doug. 'I've got a few things to do.'

'Let's go up to the Darwin High School,' said Bernie. 'That's where they're mustering all the women and children for evacuation. I think we might find something to do.'

Decisions had already been made to evacuate the women and children because of the fear of disease and the belief that the city just couldn't cope until order was restored.

Darwin High School had stood up to the cyclone pretty well, but it had not been built to cope with hundreds of anxious mothers and little children. They were milling around, trying to find out what would happen. There were police officers looking harassed as they tried to create order out of the confusion. This kind of situation was not included in their training manual.

We approached a policeman who appeared to be in charge, explained who we were and asked what we could do to help.

The policeman looked flustered for a moment and I thought he might politely tell us to keep out of the way. Under the kind of pressure he was experiencing, it is hard to think straight. Then suddenly a thought struck him.

'A lot of these mothers have babies with them. They might have to wait a while, and there's no way for them to feed the kids.'

We went up to a couple of the mothers who were carrying babies and asked them how they were going to feed them.

'I think most of us brought bottles and tins of baby food,' said one. 'But there's nowhere to heat them.'

Bernie thought quickly.

'You make a fireplace,' he told me, 'and I'll go and get something for heating the water.'

It was not hard to find bricks and timber in a city strewn with wreckage. I set up the fireplace and waited for Bernie to return. Some of the mothers came up and politely asked me when it would be ready. After a couple of hours I began to think Bernie had had an accident.

I had just about exhausted my supply of plausible excuses when Bernie appeared. He was holding two large galvanised cans and his smile was weary but triumphant. We started the fire, filled the cans and soon had the water hot enough for the bottles and tins of babies' food.

'What kept you?' I asked when we were able to talk.

'I should have woken up earlier. With the water cut off and people queuing up at the pipeline, everything from a jam tin to a bath has been commandeered for holding water. I couldn't find anything, anywhere.' He gave a sheepish grin. 'I was even prepared to borrow something out of one of the empty shops, but someone had been there before me.'

'So where did you get these beautiful cans?'

'Out of a couple of toilets,' Bernie was looking very smug.

'You did what?'

'I went down to our store. There's a lot of stuff there waiting to be shipped out to Arnhem Land. It's a hell of a mess, sugar and stuff spilled all over the floor and wet into the bargain. At first I couldn't see anything we could use, and I was just about to give up. Then I noticed two of those port-a-loos. They have cans inside, and since they were brand new they would be just the thing.'

I looked across to where the mothers were blissfully heating their babies' bottles and tins.

'You mean to tell me...?'

'Yes. But untouched by human hands, or whatever.'

'Still,' I said, 'I don't think we'll tell the mothers.'

We left the high school and went back to the house where all our mates had gathered. It had been another long and weary day. Everyone was affected with emotional and physical exhaustion, coupled with the oppressiveness of the humidity which hung over the city like a pall.

We all tried to be light-hearted and joked about the funny situations which always emerge when people have to face the unexpected. But underneath was the anxiety and uncertainty of what might happen in the days ahead.

People had begun to leave the city by the thousands. We would hear later of the little towns south of Darwin, such as Katherine and Kununurra, which were flooded with refugees on the road. They were frightened and bewildered. Many of them had no idea where they were going. Their one aim was to put Darwin as far behind them as possible. People in Katherine and Kununurra and places further south did what they could to provide food and shelter.

In the meantime the organised evacuation by air was beginning to escalate. At first the RAAF bore the brunt of the task. The ponderous Hercules transport planes packed in people in a way that was never envisaged by their designers. The air crews made the long trips to the cities in the south and back again, sometimes without a break. The evacuees often had no idea where they were going until the last moment. Everything was driven by the desire to get as many people as possible out of Darwin.

Rumour was rife in the city and at times I wondered whether someone knew the whole of what was happening.

'Does anyone have a grasp on the whole situation?' I asked Doug, at the end of the third day.

'You know they sent Major General Alan Stretton up to take charge. Well, I think he's got a fair grip on the situation now. There are a few blokes working with him. Bill McMillan, the chief of police, Ray McHenry, who's the head of the public service and Hedley Beare, the Director of Education. They're all capable fellows.'

We had come back to Doug's place at the end of another long day.

'I'm afraid it's barramundi or steak again,' he said wearily.

'It could be curried cardboard, for all the interest I've got in food,' I told him.

After we had eaten, Doug suggested we go up to the control centre and see how Alan Stretton and the others were faring. 'They might like a bit of company for a while.'

We drove slowly through the dark deserted streets of the city. It was Saturday night and normally Darwin would be alive. Tonight it was at its oppressive worst. The build-up of the humidity signalled the possibility of a thunderous storm, but it might stay that way for days. You never can tell in the north.

Police headquarters, where the control centre had been set up, was in darkness. We stumbled up a staircase and found our way to the control room, which was palely lit by the eerie glow of the radio and other equipment. It made the gaunt and strained faces of the occupants look sallow.

The first few days of the aftermath of Cyclone Tracy were already beginning to take their toll of those who had to make the decisions. What was decided would affect the lives of thousands of people for years to come. But decisions had to be made.

Heroes too are human. Doug was a warm-hearted person who knew most of the men in the room. He spoke words of encouragement and told stories of some of the funnier things which were happening. It was a break from the seriousness of all they had become accustomed to hearing. After a while, we left them to return to their brooding watch over the stricken city.

As we stumbled down the dark dank stairwell and back into the night, we heard the sound of rain falling. The drops were slow, heavy and ominous.

'We're going to get bucketed,' said Doug as we hurried towards the car. 'There's a lot of people out there tonight who won't be welcoming that sound.'

We drove back to the house and stumbled into its darkened interior. Our torches cast ghostly shadows around the empty room. The sweat and dust and fatigue of the past few days hung heavily on us. I stood for a moment, letting the despondency take possession of me and hoping that a good night's sleep would help.

'What I'd really like right now,' I said, thinking aloud, 'is a nice warm shower.'

Doug suddenly leapt up.

'Why waste the one outside?' he shouted and disappeared out of the room. In a moment he was back with two cakes of soap and some towels. 'Here you are! Strip off and come outside.'

I hesitated for a split second and then tore off my clothes and followed him out into the night. The rain was pouring down. As fast as we soaped ourselves the rain sluiced it away and, with it, our weariness and despondency. We heard later that someone driving along the main street of Darwin that night swears that he saw two naked figures dancing around in the main street, singing 'Onward, Christian Soldiers' at the top of their voices.

Sometimes you discover your sanity in a moment of insanity. And in the days that followed, sanity began to return to Darwin.

'How's Brian?' I asked Bernie, referring to our mate who had lost his family.

'I think he's coming out of the shock. Yesterday he came into the office, sat down at his desk and started work. I think he'll be all right.'

People began to come out of the shock in all sorts of extraordinary ways. One of our team in Darwin found a motor mower somewhere and, without asking, simply went around the streets, mowing the grass. People who were busy trying to restore order to their houses appreciated it.

'We think you ought to go back to Sydney.'

I looked up and saw Bernie and Doug standing over me with resolute firmness.

'But you need every pair of hands you can get,' I protested.

'You'll be more use to us back in Sydney. We're going to need massive assistance to get everything back in order. You've seen the mess and you understand what we'll be asking for. Go back and get an appeal started. We'll also need lighting plants, building supplies and experienced builders.'

There was no answer to that. The next day they drove me to the airport.

'How am I going to get on a plane?' I asked.

'You'll find a way,' said Doug as they drove off and left me.

I turned and walked towards the terminal. It was thronged with families, husbands bringing their wives and children for evacuation. Their faces were filled with anxiety and there were many tears as they were about to be separated, not knowing when they would be reunited.

There were queues being formed as people lined up to be told what would happen. I didn't think I was likely to get a seat on a plane, so I walked over towards the RAAF base. It, too, was a hive of activity. Hercules and other planes were lined up outside, being prepared for the long flight south. The briefing room was filled with air crews, snatching a respite before receiving their instructions.

I stood and watched the scene and then looked back across at the terminal, where the evacuees were gathered. My hopes of getting out of Darwin began to fade.

'You want to get out?'

I turned and looked at the man who had spoken to me. He was dressed in slacks and a bomber jacket. He had a beret on his head. It was not a uniform I recognised.

I explained to him how I had come to Darwin and why I wanted to get back to Sydney.

'I might be able to help you,' he said.

'Are you with the RAAF?' I was still puzzled by his unorthodox uniform.

'No, I'm a solicitor actually.'

He laughed. 'I came out here a few days ago to see if I could help. I noticed that the people who were being evacuated were over there and the planes were over here. There didn't seem to be any communication between the two, so I borrowed a bicycle and started to ride between the terminal and the base, finding out how many people wanted to go where. When a plane was ready, I would ride back and let the other crowd know.'

While we were chatting, the officer in command of the base came in and called all the air crews to attention.

'I want to thank all you men for the magnificent job you have done,' he began. 'As you know, the number of commercial aircraft coming into Darwin is increasing. They are much better equipped than we are to take out the women and children. In fact, I think we have reached the stage where we can now call it a day.'

My hopes of getting out with the RAAF began to sink. Then my friend with the beret spoke.

'One of the Hercules is about to return to Sydney and there's a group of people who have been waiting for a while now. Perhaps it can take them back.'

The commander paused for a moment to consider this, while I held my breath.

'All right, Tom,' he said finally. 'We'll do that. But it will be our last flight.'

Tom turned and winked at me as I shook him warmly by the hand.

I packed into the Hercules with 180 other people, mainly women and children. We sat on rows of seats that ran down the length of the interior, shoulder to shoulder and knee to knee. Despite the trauma of separation and the total uncertainty of what would happen, the people were remarkably composed.

The Hercules was not equipped to meet the natural needs of a big crowd of children. The only toilet was a canvas-surrounded unit at the back of the plane. All night long we handed the children from one person to another as they made their way to the far end of the plane and then came back again.

'I've seen you before,' I said to one freckle-faced youngster, who had decided to make a game of it. Everyone entered into the humour of the situation.

'Where are you headed?' I asked the woman sitting next to me.

'My family live in Adelaide and that's where I hope to take the children and stay for the time.'

I pointed out that we were headed for Sydney.

'Yes, I know. But that's more than half way.'

Later in the night I went up onto the flight deck. Both the pilots were young men in their twenties. They seemed unfussed by all the drama and cheerfully answered the questions people asked about the plane. I gathered they had been flying most of the week.

As Sydney approached, everyone turned and gazed out through the wide glass windows of the flight deck. Ahead we could see a twinkle of lights that grew stronger as we drew nearer. It was after midnight and the flight had been a long one, but the whole plane hummed with a subdued buzz of excitement. We stared down on Sydney, which in the black of the night looked like a star-studded sky that had been turned upside down. It was a welcome contrast to the sight of Darwin on that first day.

The Hercules landed and the evacuees disembarked. They stumbled down the steps at the rear of the plane, clutching their children and blinking their eyes in the unaccustomed glare of the airport's floodlights. Helping hands guided them to the terminal building.

I joined the crew in sweeping out the plane. Then we climbed back on board, closed the tailgate and took off for the short flight to Richmond base. I stood on the flight deck while the Hercules landed with an amazing lightness for such a large plane.

When I stepped off at Richmond, it was 2.30 a.m. on New Year's Day. I thanked the crew for the trip. We shook hands all round.

I turned and walked into the dark towards the park where a week ago I had left my car.

It seemed like a year and Darwin was another world.

10

When Things Go Wrong

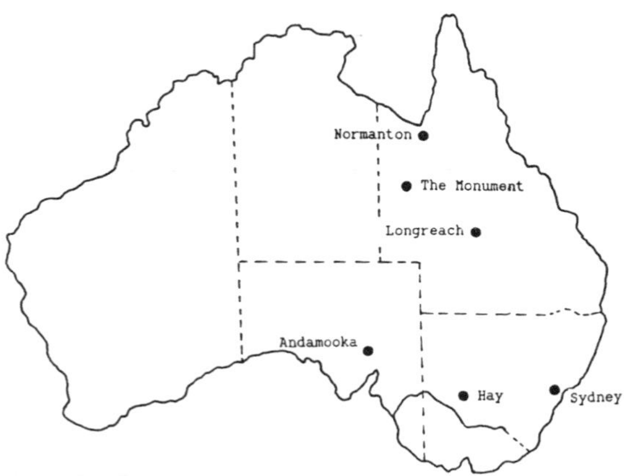

I came to dread January. It was the time of the year when things went wrong.

The weather had a lot to do with it. January was the cyclone season in the north, as many a battered community will bear witness. Tracey was the biggest, but not the only cyclone that sent us scurrying north in the anxious aftermath of destruction.

The wind was not the only destroyer. Heavy rains, followed by widespread floods, were a consistent feature of the January syndrome in the north. In fact the destruction caused by flooding was more damaging than that caused by cyclones, and far more widespread.

Take Normanton, for instance. A quaint old relic of history, just south of the Gulf of Carpentaria in North Queensland, it was spawned in the gold-rush days of the last century. Some of its buildings, like the Burns Philp store, the railway station and one of the banks, are as historically interesting as any in the outback. Others, like the pub which was painted a bright purple, are not for the National Heritage Register.

Normanton was the scene of one of my first January disasters. It was the beginning of 1974 and we were about to open a new children's hostel in the town. It was an important venture. There were many children in the isolated parts of North Queensland, who had little hope of getting any education. Normanton had a good high school. With the new hostel, bush children had a place to stay while they attended school.

We appointed two sisters to run it. Lorie and Rhona were an extraordinary couple. Both were over sixty and bush children held no fears for them; Lorie had once been in charge of a children's hostel at Halls Creek in the Kimberley. They went up to Normanton in early January, and one of my colleagues went with them to help get things underway.

The first indication of trouble came when my colleague Harry, rang me.

'We've reached Normanton,' he said, 'but things aren't looking too good.'

'What's the trouble?'

'We're having heavy rain and it's creating havoc. I've just flown over the surrounding country. It's under water all the way up to Karumba on the Gulf. You'd think the whole coastline had been moved eighty kilometres south.'

'How close is it to you?'

'Too close. The way the water's rising, it may reach the edge of the town tomorrow.'

'You'd better talk to Lorie and Rhona about leaving. I don't want them having to fight a flood.'

'Tell them yourself,' was Harry's retort. 'They have no intention of leaving.'

'Well, ring me again tomorrow.' I put the phone down.

The hostel was on rising ground and there was a possibility that it might escape the floods. But I wasn't excited about the prospect. In any case, being cut off by the floods in that isolated region presented serious difficulties.

Next day, Harry rang again.

'It's still rising,' he reported. 'They're talking about evacuating the whole town. I've spoken to Lorie and Rhona, but you know them. At the moment they are busy organising an emergency kitchen.'

The next news I heard was that Normanton had been cut off by the flood. The swift running waters had swept through some of the houses and submerged others. Then I received another phone call.

'We're in Cairns,' said Harry cheerfully. 'The whole town has been evacuated. They flew us out yesterday.'

It was a month before people were able to return to Normanton. I hurried up there as quickly as I could.

'There's no damage to the hostel,' said Lorie, when I arrived. 'We were lucky. Most of the town was affected and some of the houses were saturated. But the water only reached the corner of the hostel.'

'I believe you two were the last women to leave Normanton,' I remarked with admiration.

'No we weren't,' replied Lorie. 'Mrs L was the last person onto the plane.'

'Yes, but that woman is late for everything,' chimed in Rhona. 'She got to the airport and then decided she needed to go to the toilet. They showed her the one on the Hercules, but she said she couldn't possibly use that primitive thing. Then they suggested the one at the airport, but she wouldn't use that either. She demanded to be taken back to her own home. The rest of us sat in the plane and cooled our heels until they brought her back.'

I asked about the children in the hostel.

'We've got twelve in and there are more to come,' said Rhona, who by this time had given up trying to listen to the conversation from the laundry while Lorie and I peeled potatoes in the kitchen. 'It's been pretty tough on them. They've missed a whole month of school.'

Missing school for long periods at a time was not an uncommon experience for bush children. Floods and the financial problems of their parents were just two of the frustrations that kept them away. I thought of all the city parents who complained if their children missed half a day. Bush children's schooling often proceeded in fits and starts. They became so frustrated that, by the time they were teenagers, they had lost interest and motivation.

Still, despite the disastrous beginning, 1974 at the Normanton hostel was a good year for some bush children who otherwise might have had no school at all.

I had not long been back in Sydney from North Queensland, when Doug, my business manager, strode into my office with great agitation.

'Andamooka's flooded,' he said.

'Come off it! It's the driest place in Australia.'

That was true. The region around Lake Torrance in central South Australia, has the lowest rainfall of any place on the continent.

'Well,' Doug said, brushing aside my protestations, 'I've had confirmation from several sources, including our nurses and the flying doctor. Both airstrips are under water and the only road in from Woomera is impassable.'

I tried to envisage the hard barren dryness of the little opal mining community under water and found it very hard.

'Well, the hospital should be all right,' I said, trying to find something encouraging. 'It's on the highest spot around.'

'Yes, I've spoken to the nurses. It's okay. But they've got another problem. There are people in the town on medication, and they need it all the time. Some of them are almost out of their supplies.'

'Ring the flying doctor in Port Augusta,' I suggested, 'and see if he has any bright ideas.'

Doug came back a little later.

'The doctor said he had an idea that might work. He'll ring us back tomorrow.'

The flying doctor was back on the phone the next day, and cheerfully announced that he'd solved the problem.

'Were you able to land?' I asked.

'No, the strip will be out for a while yet, but I borrowed my little girl's doll, which has a screw-off head. I filled the inside with tablets and put the head back on again. The I made a parachute out of an old sheet and tied the doll to it. We flew over the airstrip and I chucked it out.'

His enjoyment came over the phone as he laughed.

'I looked down and saw the nurses standing open mouthed as the doll came floating down. We've been in contact with them and everything is okay.'

I breathed a sight of relief. However, any illusions I may have had that the drama was over at Andamooka, were quickly shattered a few days later, when there was another urgent message from the nurses.

'We've just had a woman brought to us in labour. The mother has some complications. She's going to need a caesarean.'

'Then you'll have to fly her out.'

'The strip's still unusable and the road's still under water.' Suddenly, her voice brightened. 'I know. They have a helicopter at Woomera. Maybe they can get her out.'

'Well, ring the flying doctor and see what he thinks.'

Helicopters come in all shapes and sizes and I had an uneasy feeling that the one at Woomera was not all that big.

Several hours later, the phone rang again.

'We've done it!' The nurse from Andamooka sounded triumphant. 'The helicopter landed in the backyard of the hospital.'

I had a vision of that hard stony piece of land on which the hospital was built and the helicopter throwing rocks in every direction when it landed.

'It was only a small helicopter,' the nurse went on. 'But they strapped Chris tightly and lifted off. She made it to Woomera and the flying doctor flew her to Port Augusta.'

Flooded airstrips were the flying doctor's biggest nightmare. It was particularly worrying in North Queensland, where tropical downpours put the station airstrips out of action for weeks.

I met a woman once, who lived on a property north of Julia Creek. She was out riding one day when her horse stumbled and fell. The woman injured her back seriously and was in great pain. When she didn't return, they went out looking for her and found her lying on the ground, unable to move. She had to endure the agonising trip back to the homestead on the back of a truck.

It was the wet season and the station strip was unusable. So there followed another long and agonising trip from the station to Julia Creek. Here she was transferred into a road ambulance and driven 260 kilometres to the base hospital at Mount Isa. Her condition was so serious that she was immediately placed on a plane and flown 2000 kilometres to Brisbane.

I attended a meeting in Julia Creek not long after that incident. The people were agitating for a helicopter to be stationed somewhere in the region and I could understand their concern.

Despite the magnificence of the Flying Doctor service, not every medical emergency in the outback can be met in the routine way. Sometimes, when things go wrong, desperate measures are necessary.

Desperate measures were needed to help Bill and Kerry. They had just been married in Sydney and were on their way to work in Alice Springs. The long drive was to be their honeymoon. The route would take them north to Mount Isa, then west to Tennant Creek. From there, they would head south down the Stuart Highway, to Alice Springs. There were stretches of road on the way north to Mount Isa where the bitumen was narrow and the edges jagged and broken. It was on such a stretch that Bill and Kerry's honeymoon came to an abrupt end.

They were travelling towards Longreach. It was late in the afternoon and a fine misty rain made visibility very poor. Without warning, the wheels of their car slipped off the crumbled edge of the road. Despite Bill's efforts, the car went wildly out of control and ended up on its side with Kerry pinned up against the passenger-side door.

As Bill struggled to get his door opened, he realised that his leg was not functioning. Kerry, in the meantime, was beginning to feel a painful throb in her elbow. Neither of them could move. As they lay there waiting for someone to come by and discover them, the pain of their injuries began to mount.

Traffic along that outback highway was not exactly bumper to bumper. Fortunately, someone eventually stopped and came to their rescue. An ambulance came out from Longreach and took them back to the hospital. Bill's leg was broken and Kerry had a complex fracture to the elbow.

'You've got to get them out of there.' The voice was calm but urgent. It was Kerry's mother, who worked in a major Sydney hospital. 'I've spoken to the doctors at Longreach. They're doing everything they can. But the hospital is not equipped to deal with the complication that Kerry has. The people at my hospital say if she doesn't get proper treatment quickly she'll be affected for the rest of her life.'

Kerry's mother had already made enquiries and discovered that the flying doctor could not transport her daughter. Neither was the small commuter aircraft that flew from Longreach equipped to cope with the two casualties. We had to think of something else.

When things go wrong in the bush, it is generally the bushies themselves who are most likely to come up with the answers. So I picked up the phone and rang Leonie. She lived on a property in southern New South Wales and also happened to be an excellent pilot with a lot of experience flying in the bush.

I explained the situation to her. There was a long silence.

'The Moonie's too small,' she said slowly, referring to her own little single-engine plane. 'We might do it in a 210.' This was a larger, faster, single-engine aircraft, much favoured by the people in the bush. 'I think I might be able to borrow one from Albury. If you hang about for a few hours, I'll get back to you.'

As I sat and waited for the call, I began to have serious misgivings about the whole project. Longreach was about 1500 kilometres from Sydney. Leonie's property was about the same distance but in the opposite direction. Albury was not exactly over the back fence, either. It would require a flight of 3000 kilometres from Leonie's property to pick up the patients at Longreach and bring them to Sydney.

But the real problem was how to accommodate a man with a broken leg and a woman with a badly damaged elbow in the 210 for a flight of 1500 kilometres. I was still worrying over this when the phone rang.

'I've got it!' said Leonie. There was a note of excitement in her voice. 'We've modified the doors of the 210 so the chap with the broken leg can be lifted in. The hospital people here are helping me fit up the plane to carry them. I'll leave for Longreach first thing in the morning.'

'I want to come and help,' I said, recognising the responsibility I was asking of her.

'There's no room. We've taken out all the seats except the pilot's. The guy with the broken leg will lie stretched out on the floor and the woman will have a special sling to help her arm. She'll sit on the floor beside him.'

'Well, what can I do?' I asked, feeling a little redundant.

'You can stay in Sydney and deal with Air Traffic Control. I'm going to need a lot of help. Oh, and also, pray for a fine day.'

Next day, late in the afternoon, I had a report from Longreach to say that Leonie had arrived safely.

'It's causing a bit of a stir up here,' my informant said with a chuckle. 'Bill and Kerry didn't tell the doctors what was planned because they reckoned the doctors wouldn't have a bar of it. As a matter of fact, they were very resistant to the idea. When they saw the plane, they nearly had a fit.'

'So what's going to happen?'

'Well, that Kerry is a very determined young woman. They've agreed to let them go. They've fixed up something on the wall of the plane that will support her arm. All being well, they'll fly out tomorrow morning.'

'Tell Leonie that I've fixed it with Air Traffic Control. She can fly straight through the restricted airspace over Richmond RAAF base. They don't think she's likely to drop a bomb on them.'

The next morning, I had another ring from Longreach.

'They've taken off,' said my mate. 'We got Bill in through the cargo doors. They both looked pretty cheerful, but I'm glad it's Leonie flying and not me.'

'How's the weather up there?' I asked, knowing the importance of clear skies both for the pilot and for the patients' comfort.

'It's not bad.'

I put down the phone and proceeded to cope with the agonies of waiting. After a while, I rang Air Traffic Control.

'She seems to be coming along nicely,' they told me. 'Although there's a bit of rough weather in the area. She might have to fly around it.'

That was all we needed. It would make for a longer flight and perhaps the need to put down somewhere, if the storm became too turbulent. According to the plan, the 210 was due to arrive in Sydney between one and two in the afternoon. I rang Air Traffic Control around midday.

'They're just passing over Richmond base now,' they reported cheerfully. 'They should be landing in about fifteen minutes.'

I leapt into my car and drove to the airport. There was no way I would make it in fifteen minutes, but I tried! It was Sunday. The gates into the light aircraft park at Mascot were wide open so I drove straight in. There was Leonie, calmly tying down the 210 as if she had just been for a joy ride. There was no-one else in sight.

'What kept you?' she asked with a mocking grin. 'The ambulance was waiting when I arrived. Bill and Kerry are probably already in hospital.'

'So where did you get the jet engines that brought you here so fast?' I asked, feeling a bit left out of all the drama.

'Flying conditions were fine for a while,' explained Leonie. 'Then I saw this big storm building up right on the flight path. I couldn't risk going through it with Bill and Kerry, so I decided to fly round it.' She smiled softly. I suddenly realised how tired she was. 'Someone was on our side. We caught a terrific tail wind. I didn't even have to land for refuelling. We sailed home.'

Leonie went off for some well-earned sleep and the next day flew the 210 back to Albury.

As I said, if you have a problem in the bush, ask the bushies for help.

You can't blame the weather for all the disasters that happen in the outback. Errors of human judgment and plain stupidity contribute their fair share. The people of the outback are particularly vulnerable when decisions which affect them are made by those in authority in centres that are far removed from them.

Someone decides to close down an air service that is the only lifeline between a little outback community and the outside world. It's like severing an artery.

Someone decides to put up the price of petrol, or remove a subsidy. If it happens during a drought, it can break the will to battle on. The vice-like grip of isolation finally crushes the human spirit.

Some of the communities of the outback which are affected in this way belong to the mining industry. The public face of the mining industry is one of power and economic might. It's hard to equate this with the plight of a small isolated community. But when a mine is closed in the outback, the effect on the people who live and work there can be devastating.

It happened to the people in a tiny outback town called The Monument.

The road south from Mount Isa gives no promise of exciting things ahead. When John Flynn and Alf Traeger travelled through that country, installing the first pedal wireless sets on isolated cattle stations, they had little idea of what lay under the ground beneath their wheels. But then most of the country in the outback where rich discoveries have been made shows little of interest.

There is a turn-off from the main road south from Mount Isa, near a little settlement called Dajarra. The track winds through the bush, crossing several dry creek beds. Then ahead of you is an extraordinary sight.

It is a small cone-shaped mountain, standing alone on an otherwise undistinguished landscape. At the pinnacle of the mountain, a large rock outcrop stands erect. Since it looks for all the world like a massive monument, it required little imagination to name the place The Monument.

The Monument sprang into public prominence with the discovery of an extensive seam of phosphate rock. Phosphate had been the Achilles heel of the rural industries of Australia. During the Second World War the Japanese sealed off the Pacific islands of Nauru and Christmas Island, the major source of phosphate for Australia. It exposed our considerable vulnerability. The discovery of huge deposits of phosphate rock in North Queensland gave rise to the hope that Australia would become self-sufficient.

The phosphate discovery came at a time when 'feeding the world's hungry' was a popular catchcry. The politicians were not slow to recognise the significance and made much of it.

However, the mining company concerned was more conservative. 'It's not going to be a big operation,' they told me, when we were negotiating the possibility of conducting their hospital. 'Maybe thirty to thirty-five houses to begin with and of course the single men's quarters.'

'That doesn't worry us,' I replied. 'It's about what Oodnadatta had when John Flynn built their first hospital. Anyway, the government seems pretty pleased about the deal.'

Tailboard medicine. Sister Maisie Ross visiting an Aboriginal camp outside Halls Creek.

Care for the aged. Sister Nancy Cole with friends at Fossil Downs station, West Kimberley.

Nature's memorial. Rock outcrop above the phosphate mining town called The Monument in north Queensland.

'I'm afraid the government is not going to be as charitable as we had hoped. In fact they're going to bleed us white. We'll have to build our own branch railway line to the main Mount Isa–Townsville line and then hand it over to the government. They'll charge us exorbitant rates for its use.'

Despite the reservations, the company went ahead with the development. They built us a little hospital and we recruited two experienced women to run it. I drove with them from Mount Isa down to The Monument. They had brought their own four-wheel drive and caravan.

'We'll live in the caravan and conduct the clinic there until the hospital's finished,' said Virginia. 'That way you start off roughing it with everyone else, and you get to know the people.'

I helped them set up the caravan and then wandered through the beginnings of the new town, chatting to people.

'I'm glad the nurses are here,' said one of the women I met. 'We're having a plague of brown snakes and death adders. With the children running all over the place, we need the security.'

Jock, the town manager, told me with some pride that although it would only be a little town, the company had planned it so that it would be attractive and have plenty of amenities. He took me for a tour of the mining operations.

'We've located the town about fourteen kilometres from the mine,' he explained. 'There won't be the noise and the dust that some of the old mining towns have.'

We came to the edge of the open cut from which the phosphate rock was being excavated.

'It's not difficult stuff to get out,' said Jack, the mine manager.

I looked down into the big pit. The rich red topsoil was in stark contrast to the glaring white of the phosphate rock underneath. Bulldozers were burrowing away at the bottom and ore carriers were crawling slowly along the tracks which had been cut into the sides of the mine.

'It's a pretty simple operation, as you can see. We dig it out, put it through the crusher and then load it onto the train.' He pointed to the crushing plant which towered over the open cut.

We walked towards the railway, where a line of open trucks waited for their turn to be filled. The conveyor belt poured the crushed white rock into them and quickly satisfied their appetite.

'The government is swallowing our profits as quickly as that truck is being filled,' said Jack.

'I hope it doesn't cripple the operation,' I replied.

Jack looked gloomy. 'We'll see,' was his only comment.

I returned to The Monument a few months later. Despite the government's high rhetoric about feeding the world's hungry, the mine was still struggling to break even. In fact it was losing a lot of money.

'There was a rail strike recently,' said Jack, 'and the fellows at Mount Isa Mines used road trucks to carry their ore. They saved a packet. But unfortunately the strike ended and they had to go back to the rail.'

The mine continued to operate, although from time to time we heard rumours that it might be closed. Two years after it began, I received a phone call from Kath, the nurse then working at the hospital.

'They've just told us the mine is going to close.' There was a note of sadness in her voice.

I was not surprised. The high cost of transport, plus the government's unwillingness to subsidise the mine, meant the company was simply pouring money down the drain. I flew up to The Monument to see how the people were taking it.

'It's pretty devastating,' said Kath. 'In the past two years, the people have become a close community. I really enjoy it here and I hate to see them having to face this anxiety.'

I went off to the mine to see Jack and asked him how long it would be before the mine closed. He laughed ruefully.

'Well, it's a bit of a slow death. There's a lot to do. We're going to mothball the mine and the town. That will take a bit of time.' Then he turned to me with that frank open honesty I have often found in the miners. 'As far as the hospital is concerned, we appreciate what you have done. But if you've got better things to do elsewhere, then we'll understand if you want to leave now.'

'I didn't come up here for that reason,' I replied. 'Kath wants to stay. She's a pretty responsible person and I think it might be helpful if she does.'

Jack's face showed a glimmer of relief. 'Yes, well I know the people here think a lot of her. She's a bit like a mother to many of them, even to me.'

So Kath stayed to the end and mothballed the hospital along with the town and the mine.

'Somehow,' she said to me as we were leaving, 'I don't think we've seen the end of this place.'

It took three years for her prophecy to be fulfilled. By this time, Western Mining Corporation had taken over the phosphate company and decided to give it another go. One of our senior nurses, Beth, went up to take the mothballs out of the hospital.

'It's in remarkably good condition,' she reported over the phone. 'I've contacted Ian, the flying doctor. He'll be able to come down on regular visits. Mount Isa Hospital has been good and will supply us with medications.'

'I've been in touch with Kath and she's prepared to go back again,' I said.

Beth responded enthusiastically. 'There are a few people who were here before and they'll be glad to see her.'

A few weeks later, I rang Kath at the hospital and asked her how she was settling in.

'Great!' she replied, but with a touch of irony. 'I had a wonderful welcome. All the children are down with chicken pox!'

My next trip to The Monument was typical of the unexpected situations that happen in the outback. I was in Julia Creek, a small town on the road and rail link between Townsville and Mount Isa. We were holding a conference to discuss some of the urgent medical concerns in the region.

Kath had come up from The Monument. When the conference was ended, we were to fly back with Bob, one of our padres, who was based at Cloncurry. It was late afternoon when we finally got off the ground. Bob had a reputation for last-minute departures. We flew south towards The Monument and the sun was declining in the death throes of the day.

'We're not going to make it before that sun disappears,' I said to Bob.

'We don't have landing lights at The Monument strip,' added Kath calmly.

Bob looked anxiously out of the cockpit. 'See if you can spot a suitable place to land,' he said.

The desert below us was not about to be helpful. We scanned the country in every direction. Eventually I pointed to a short stretch that had been scraped clear by a bulldozer.

'I'm not sure if it's long enough,' commented Bob.

'I'm not sure it's wide enough,' said Kath from the back seat.

'Well, it's all we've got,' said Bob, shortly, 'so here we go.'

The plane wavered through the air, losing height all too quickly. We waited to hear the wheels bump into the earth. The plane slewed a bit as the soft sand gripped the wheels, but Bob kept it under control and was able to stop before the graded earth gave out. Bob shut off the engine. We sat in silence for a few moments.

'There was a funny noise while we were on the way down,' I said, 'a clicking noise. For a while I thought something was breaking up.'

'That clicking noise,' said Kath, with great dignity, 'was my rosary beads.'

'We'd better let someone know where we are,' said Bob. 'Air Traffic Control doesn't take kindly to planes that don't report in.'

Radio contact from the ground to Mount Isa was not easy, but eventually we were able to make ourselves heard.

'You're very faint,' said the controller at Mount Isa. 'Where are you?'

Bob looked embarrassed and then said he wasn't sure.

'Well, what colour is the soil? That might give us a clue.'

I jumped out of the plane. It was pitch black. I scooped up a handful of soil and jumped back in again. The light from the instrument panel was very weak. We had a fierce argument whether the soil was black or red, while the controller at Mount Isa waited patiently.

'We think it's red,' said Bob, while I shook my head vigorously.

'It doesn't matter,' replied the controller. 'We think we've worked out where you are. We've contacted The Monument and they'll be out to collect you in the morning. Have a good night's sleep.'

It was a hollow mockery. Outside the desert was beginning to chill and we sat for a while in silence in our light summer clothes.

'What's that glow on the horizon?' said Kath suddenly.

We looked and saw a faint glow in the distance.

'It could be the town lights at The Monument,' said Bob, with renewed confidence. 'If so, we're not far away after all.'

We looked at the glow as if it were the light at the end of the tunnel. Suddenly, it went out.

'That's funny,' said Kath. 'They never turn the town lights off.'

The sense of security the glow had given us dimmed away very quickly. We sat and talked for a while. The cold air began to penetrate the thin veneer of our clothing.

'There's another light!' shouted Bob. 'And this one's moving. I'll turn on the beacon.'

The red revolving light on top of the plane's cabin began flashing. At first it seemed to shine with brave strength, but after a while it lost its first radiance. To make matters worse, the moving light on the horizon disappeared and did not return. We returned to our gloomy contemplation of the night ahead.

Then, suddenly, the bright headlights of a vehicle shone out from the darkness, just a short distance away. We jumped out and began waving frantically. The vehicle kept coming and pulled up in front of us.

'Thought we'd better rescue you tonight,' came the cheerful voice of the mine manager. 'It's going to be a cold one.'

'We saw a glow in the sky,' I said to Jack as we drove back. 'We thought it was the town lights, but suddenly they went off.'

'That would have been the tennis court lights. They're brighter than the town lights.'

We got back to the town and a hot shower.

The Monument mine kept going for a few more years and eventually closed down. By that time I was not connected with the work, but I felt a deep sense of sadness when I heard the news. The Monument might not be one of the roaring success stories of mining in Australia, but it is a story of how people in the outback handle situations when things to wrong. The people in the little community faced the normal problems of health, personal relations, education and job security. However, isolation brings an added dimension of anxiety and suffering that city people do not know.

The closure of the mines at The Monument and in many similar places in the outback meant the complete upheaval of personal, family and community life, as well as employment.

It's hard to watch a town die. But it's a privilege to work with people who fight the inevitable to the end and then accept the inevitable with dignity.

11

No Neighbours to Speak Of

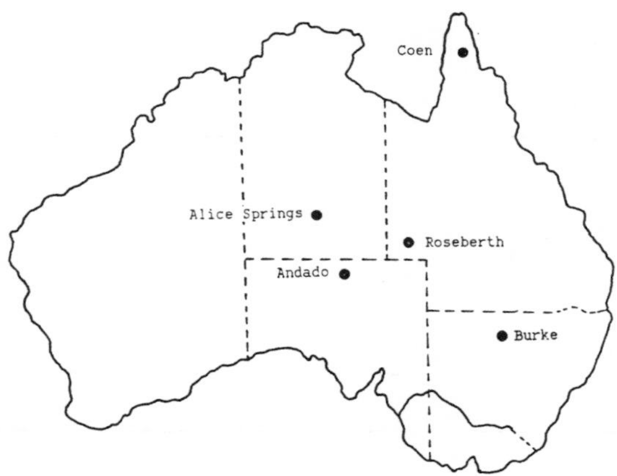

Les wasn't looking his best. He slipped furtively past the door and cast a guilty glance into the room. His hair was wild, his shirt stained with dirt and grease and the seam of one leg of his trousers was torn open from the knee down. I think he hoped we hadn't seen him, so I pretended I hadn't. But his wife had seen him.

'Les has been out fixing a windmill,' she said. 'It looks as if he has struck a bit of trouble. He'll be out when he's cleaned up.'

The small sitting room was painted a pale green and the lounge suite was covered with floral chintz. There was a crystal cabinet topped with family photographs and two framed photos of racehorses hanging on the wall. It was an artificial room that seemed to be the product of some spurious social necessity. It was quite out of character with the rest of the house and the people who lived in it.

Flying across the Australian outback on a hot summer's afternoon is a tiring exercise and the boredom of the land beneath doesn't help. Still, you have to keep your wits about you. Landmarks are few and far between. If you miss one of them, it's a long way to the next and a lot can happen in between. There was an hour of daylight left and already the haze on the horizon was making visibility difficult.

Suddenly the pilot said, 'Les asked me to drop in and see him next time I was passing. He's had a bit of trouble lately, so we'll probably stay the night.'

Shortly after, I saw the long strip of graded earth that marked the station airstrip. Just beyond it the last rays of sun glinted on the iron roof of the homestead. The pilot made a final wheel around the sky and came in to land. We taxied towards a large shed at one end. I noted a single-engine plane standing outside. I looked again. Someone had taken away the propeller and punched its nose.

'Les was warming it up,' said the pilot as we taxied past it. 'The brakes weren't on properly and it ran into the hangar. He may have to truck it out and get it fixed. It'll take a while, which is a nuisance. He needs it now because it's the mustering season.'

I looked out at the hazy horizons which blended into the mellowing sky. We were a long way from people and places which fixed up aeroplanes. A small boy in shorts and a yellow sunhat ran out to meet us.

'Gooday, Tim,' shouted the pilot. Then turned to me. 'He's the only one at home now. His brother and sister have gone away to school.'

'Whom does he play with?' I asked, looking down at the five-year-old boy.

'I don't think there's another kid within a couple of hundred kilometres.'

We walked across to the homestead. It was a typical bush building, surrounded with verandahs. In front of it was a small fenced-in garden where someone was performing miracles of growth. A pleasant middle-aged woman came out to meet us. The harsh climate had beaten but not bowed her face. This was Les's wife, Val. I complimented her on the flowers which were growing in the garden. She smiled.

'They take a lot of water,' she said. 'And often we can't spare it. The dust storms play havoc. But it's worth it.'

I looked beyond the fence to the rust-brown soil that stretched in unbroken monotony to the gold of the setting sun. Val's garden was her last line of defence against the desolation of the desert. If the garden went, then Val might also go.

About an hour after Les had made his brief appearance, he reappeared, showered, shaved and clean-clothed. Val thought it was time we ate. We left the formality of the little sitting room and came into a much larger space. Obviously this was where the family usually lived. We sat down and Val began to serve the dinner. It was vegetable soup and roast beef served in generous quantities.

The conversation began politely. It was not every day an outback family had two parsons to dinner. The two children away at school were staying at a hostel which we ran. I asked Les and Val if they were happy with the way their children were treated. They said the children were fine.

I told them a few stories about children in some of our other hostels, like the boy who forgot to take his pet frog out of the pocket of his shorts and how the hostel mother nearly died of fright when the frog suddenly jumped out of the washing machine. Gradually we all began to relax. When dinner was finished we didn't go back into the sitting room but simply sat in some of the comfortable old armchairs scattered around the family room.

I can't remember much of our conversation but we talked late into the night. Like most outback men, Les had a capacity to fire questions and then sit back and listen. I came to understand that provocation can be a powerful weapon in the search for the truth. It was also obvious that when you have one or two visitors in the course of six months, you make the most of sharpening your wits.

The only time Les showed any real emotion was when he talked about the Aborigines. He expressed opinions that I heard often around the outback. The drink was destroying them and the government was spending too much money on projects that were worthless.

We left early next morning and Les walked with us to our plane. As we passed his hangar, I expressed sympathy for the damage to his plane and hoped he would be able to get it fixed soon.

'Bloody stupid thing to do,' replied Les. 'All this space around,' sweeping his arm to the far horizons, 'and I had to hit the hangar wall.'

I had the feeling that we had come at the right time. It was a long way to the nearest pub where he might have been able to let off steam to a few mates.

The next time I saw Les and Val, I almost didn't recognise them. It was at a race meeting. I was standing on a slight rise of ground, looking over the course, when a woman came walking towards me. She had on high-heeled shoes to which she was not accustomed and she stumbled a little on the uneven ground. She wore a bright summer frock and navy-blue gloves. One hand held her straw hat firmly on her head. The other dangled a handbag that was so new the straps hadn't had time to stretch. Her face was familiar.

Then I looked at the man walking with her. He wore pressed clean whipcord trousers, a neat fine checked shirt and a new flat crowned Akubra hat. But the clothes couldn't conceal the round reddened face. I said hello and we engaged in some catch-up conversation. Then Val said, 'It's a bit different from last year. Do you remember it?'

She was referring to the race meeting. We were attending that unique function of the outback, the annual races. For many outback people, the annual event is not only the sole race meeting of the year, it's the only social event of any kind. It's the one opportunity to meet your neighbours who might live up to 300 kilometres away.

People come in from distant pastoral stations and small settlements. They toss their swags out from the back of the truck and set up camp on the ground in the middle of the racecourse. Horse trailers indicate the families who have brought hopeful starters. There are ponies too, because on one of the days there is a gymkhana, where the children show off their riding skills.

'Yes,' I said, in answer to Val's question. 'I remember what happened.'

Last year's races had been different for a number of reasons. I looked at Val's freshly styled hair and remembered what had been the most dramatic difference.

The custom in this outback centre was to fly up a hairdresser from the nearest big town, which was 500 kilometres away. On the morning of the first day of the races, twenty excited women gathered in the lounge of the local pub to take their turn under the hair dryer. The plane arrived from the south and unloaded its passengers, who included two bookies. But there was no hairdresser.

You could imagine what that could mean to a city woman. For a woman of the outback, attending the one social occasion of the year, the absence of the hairdresser was almost the end of the world.

The women were devastated and not even their own frantic efforts to do each other's hair could make up for the loss of the luxury of having a real 'hair do'.

The bush racecourse where I met Les and Val was a flat area surrounded by hills. A rail ran down one side of the straight and a simple judge's box marked the finishing post. You could tell the course wasn't used very often. The meeting lasted two or three days and the nights were devoted to dances, one of which was the fancy dress ball.

The people of the outback feed on the events which happen at the annual races and talk about them for years afterwards.

The absent hairdresser didn't supply the only drama that year. The nurse who ran the little hospital in the town also doubled as the local vet. A few nights before the races were held, she was called out to attend a sick horse. It had been entered for the cup, which was the main event of the meeting.

In the strange way that bush people come to hear of things, everyone for miles around knew that the horse was ill and that, despite the best efforts of the nurse, it was not getting better. As the condition of the horse declined, so its value increased. The initial estimate was that the horse was worth $2000, but by the time it died its value had increased to $20 000 and it had been set to win the Melbourne Cup.

On a rise overlooking the town and the racecourse was a children's hostel and the hospital clinic. During the days of the races, everyone came up, either to have the nurse treat some complaint, or to visit their children, or just to pay a social call. Strapping young men now running their family's properties would drop in and reminisce about their schooldays. Mothers arrived with cakes and fresh clothes for their children.

The hostel was a rambling U-shaped building with floor-to-ceiling louvre windows to let the breezes through. The children's bedrooms were simply furnished and the children simply dressed.

I talked with one father and mother who had two children at the hostel. The parents owned and operated a prawn boat in the Gulf of Carpentaria. I apologised that shortage of money prevented us from making the bedrooms in the hostel brighter. The mother laughed.

'You should see their bedrooms on the boat,' she said. 'They sleep out on the deck.'

I ran into Paddy and Barbara, who lived on a station not far out of the town. Their children were currently staying at the hostel.

'We don't run this hostel for parents who are fed up with their kids,' I joked. But I knew the real reason why their children were there.

Shortly the rains would come to this region. Although Paddy and Barbara's place was only twenty kilometres out, when the creeks flooded they might just as well be 200 kilometres away. They put the children in the hostel before the rains came, so they could get to school.

The rains really isolated this place. Between the town and the airport there were fourteen creeks that crossed the road. In the wet season, all of them flooded. One of my friends used to carry a midget motor bike in his plane. He drove it from the airport and, when he came to a creek, held it over his head and waded across.

The little town was linked to the outside world by one road. There were times when the rains were so heavy and sustained that the road was impassable from the beginning of November to the end of March. Living in that kind of isolation for as long as five months made the annual race meeting a time of great importance to the people.

This year the races went well. There were the usual crop of accidents that kept the nurse busy. An elderly Aborigine fell asleep close to his fire and his blankets caught alight.

Several riders fell during the races and one sustained a broken arm. There was a fight outside the pub in which both participants drew blood. They were brought up to the hospital and one of them was bleeding profusely.

The nurse ordered me to catch the blood in a bucket. When I went to throw it out, she lifted her head from the suturing she was doing and shouted at me, 'Don't waste it! Throw it on the garden!'

The competition for prizes at the fancy dress ball was intense. One year the nurse was determined to win first prize and created the uniform of a British Army officer, complete with pith helmet, swagger stick and fierce moustache. I waited with her while she finished the evening surgery and then we went down and joined the festivities. The fancy dress parade was just about to begin when there was an urgent call for the nurse to go across the road to the pub. We hurried across and were shown into a bedroom.

The scene was bizarre. There was the publican, a big man, stretched out on the bed. He had dressed up for the ball as a busty blonde, complete with wig and heavy make-up. He had suffered a heart attack.

The nurse took one look at him and leapt up on the bed. The sight of a woman dressed up as a man straddling a man dressed up as a woman, administering mouth-to-mouth resuscitation, was too much. Fortunately the publican recovered, but it took him a long time to live it down.

The final night of the races this year was spent celebrating the twentieth anniversary of the opening of the hostel. I had brought up a film which had been made at the official opening ceremony, twenty years before. Hundreds of people packed in to see it. There were two young men present who had been the first students to live at the hostel. They had to endure a lot of ribald comments about themselves as raw-boned bush boys. The crowd made us show the film three times.

By the middle of the next morning everyone had packed up and gone home. The area around the racetrack, so recently brimming with life, returned to its customary forlorn emptiness. A solitary piece of paper swirled listlessly down the straight until it finally clung to the finishing post and stayed there for security.

In the distant homesteads, hats and handbags were being wrapped in tissue and stowed away for another twelve months. A few heads were still aching from excessive and unaccustomed festivity. But there was work to be done, windmills to be checked and fences to be mended. Another twelve months would pass before the community of this region could give expression to its common humanity. But it would be worth the wait.

Some outback neighbours have to wait a little longer than twelve months to meet each other. But again the wait is worth it.

Molly had lived on the edge of the desert for twenty years. Her husband ran a huge cattle station that could have encompassed a fair slice of the state of Victoria. To reach the homestead, you crossed the wide bed of a dry old river. From that point onwards, the hungry sands of the Simpson Desert clutched at your wheels. The old homestead stood like a fortress looking out over the emptiness of the desert as if defying an invisible enemy.

Molly's husband had been killed when he had a heart attack flying his plane, and she lost a son when he was killed driving a cattle truck to Adelaide.

She continued to work the station and I dropped in to see her one day with a film crew who were making a documentary for television. We had flown down from Alice Springs. The bright red sandhills around the homestead and the old building itself kept the film crew happy for hours.

When the filming was finished we enjoyed the customary outback hospitality. Our next stop was Birdsville, which was 450 kilometres away on the other side of the Simpson Desert. There was another station near Birdsville I wanted to visit and I asked Molly if she knew the family who lived there.

'I've heard the woman on the radio many times,' she said, laughing. 'And I've often wondered what she's like. After all she is my neighbour and she sounds quite nice.' Then she turned and fidgeted with something on the kitchen bench. 'But I suppose she's one neighbour I'll never meet.'

I reflected that 450 kilometres might be a long stroll to the back fence of an outback property, but it was not all that far in a plane.

'Throw a toothbrush in a bag,' I said. 'We'll fly over this afternoon. You can go out to the station in the morning and we'll have you back tomorrow afternoon.'

The fading light of the late afternoon throws a soft and magical glow on the face of the Simpson Desert. The sharp lines of the savage sandhills are mellowed by the sinking sun. They turn to the colour of burnt orange and the crevices between the ridges often shimmer with green if the rain has fallen. In these circumstances, the desert looks enchanting, but treachery often hides behind a lovely face.

We landed at Birdsville and stayed the night at the hospital. Unfortunately for the crew the nurses couldn't produce a medical drama for them to film, but they were prepared to bog their vehicle for a film sequence, if the crew dug it out.

Then we headed for the station homestead of Mollie's neighbour. After we had driven for a while, I pointed ahead to the homestead, which had been built on a high ridge and stood prominently on the horizon.

'Why did they build it up there?' asked one of the film crew.

'The river bed we crossed is the Diamantina River. In the wet season, all the water from North Queensland flows down past here on its way to Lake Eyre. If you want to come out here you have to use a boat.'

The crew stared across the barren desert country and looked back at me in disbelief.

'Well if you don't believe me,' I protested, 'ask the crocodiles.'

Now they were convinced I was pulling their legs. We climbed the rocky winding track up to the top of the ridge.

'Take the vehicle over to that clear spot,' I commanded. 'We can see the river from there.'

We got out of the vehicle and walked to the edge of the ridge. I pointed down to one spot on the river where a couple of crocodiles were preening themselves in the sun.

'Anyone care for a swim?' I asked.

Then we walked back to the homestead. A tall grey-haired lady with a warm smile was waiting for us.

'Phyl,' I said. 'Meet your next-door-neighbour, Molly.'

For the rest of the morning I sat and watched these two extraordinary women catch up on twenty years of their lives as neighbours. Neither time nor isolation nor the harshness of the climate had diminished the essential goodness and dignity of them both. They shared many things, including their beliefs about life. The room in which we sat was an expression of the kind of women they were. It had not been especially dressed up for the occasion. It was always ready to receive visitors, even if, as in this case, it involved a wait of twenty years.

Next morning we flew back across the Simpson Desert.

'What do you think of your neighbour?' I asked Molly.

'She's just as I thought she would be,' she replied, with firm contentment.

Isolation feeds the fears and anxieties of the people of the outback in many ways. In John Flynn's day, it was the fear of becoming ill or having an accident and being beyond the reach of medical care. The nursing outpost hospitals and the flying doctor, together with the pedal radio, put an end to much of that anxiety. But isolation still created serious problems for the people of the outback.

Clare lived on a station about 800 kilometres north-east of Perth. Before her marriage she had been a nurse in a large hospital in Perth. She was an intelligent and attractive person. One night at a party she met a man who came from the bush. He was visiting his parents, who had recently retired from the property, and attending to some business. An old friend had invited him to the party and since the opportunities for a social life in the bush were few, he went.

When he was introduced to Clare, he was reserved at first, but as the night went on Clare found that he had a sense of humour and a very perceptive mind. She was not surprised when a few weeks later, she received a letter from him. Their romance blossomed slowly at first, because of distance and the busy lives they led. She visited his property on one or two occasions and found herself wondering how she would cope with the isolation. When he asked her to marry him she had already thought about the possibility and had made up her mind that she would accept.

They had two children and she found great satisfaction in watching them grow up in the healthy life of the bush. As a former nurse she found that other women, even if they lived at a distance, would drop in for a social visit, which often turned into a counselling session. When the children ultimately had to go to Perth to school, the parting was painful, but she knew it was in their best interests.

Some years later Clare became pregnant again. She and her husband were happy because by this time the steady development of their other two children and of their life in general had given them greater confidence. In due course she went to Perth to await the birth of the baby. All went well and a fortnight later she was headed back to the station.

Then she began to notice that there was something wrong with the child. Being a nurse, she knew all about professionals becoming too obsessive about their children. Instead of rushing immediately back to Perth, she waited for a while. Had she been living in Perth she would have had no qualms about going to see a specialist. But 800 kilometres was a long way to go just to be told that you were being over anxious.

The signs and symptoms did not go away. She returned to Perth and consulted a specialist. The diagnosis was cerebral palsy. The 800 kilometres back to the station was the longest and most agonising journey of her life.

Clare and her husband and the other two children were drawn more closely together by the crisis. The baby needed constant care and attention. Neighbours who lived a great distance away began to drop in with surprising frequency to do the washing or hold the fort while Clare had a break. But it wasn't enough. The amount of attention required and the need to have access to expert advice made living on the station almost impossible for her.

Gradually Clare and her husband began to talk about something which had previously been buried in their fears. The husband was the third generation of his family to work the property and it was deeply rooted in his life. It had also become a big part of Clare's life. The prospect of selling up and leaving was almost too painful to contemplate. But the family was united in the conviction that the child should have every possible chance to grow up and develop as much as possible.

In the middle of the crisis of making a decision, Clare received an invitation to attend a conference in Burke. The last thing she wanted to do at that time was to leave her family and track right across Australia and halfway back again, but something compelled her to go.

The conference was being held by the Isolated Children's Parents Association, a group of parents from all over outback Australia who shared the desire to give their children a better opportunity for their education. ICPA had achieved the first real recognition of the serious disadvantages which outback children suffer and the conference to which Clare was invited was going to focus on the needs of isolated children who had disabilities. It was believed that there were many children who received absolutely no help because of their isolation. The ICPA wanted someone who was prepared to break the tyranny of silence and plead the cause of these children.

I was present at the conference at Burke when Clare told her story. She did so with a simple dignity, but in a way that moved many who were present to tears. When she had finished, several other parents stood up and told stories of their struggles with disabled children. Others told of neighbours in the bush who desperately needed help but could not get it without travelling thousands of kilometres to a capital city.

Several times I had heard of appeals to send children overseas for treatment and I suddenly realised that there were many children in my own country who were also distanced from the kind of help that city people took for granted. Half of Queensland, for example, seemed to be without the services of a speech pathologist, and yet speech difficulties were one of the major disabilities in the outback.

Fortunately there were a number of government people present who got the message and promised to do something about it.

ICPA was an example of how the people of the bush have to use great initiative in order to overcome isolation. It was run on an entirely voluntary basis and those who came to the conference shared the travelling expenses so that the ones who had to come the furthest were not disadvantaged.

The conferences were always well-attended and the bush people took time off to enjoy some rare socialising. The conference at Burke happened to coincide with the wedding of Prince Charles and Princess Diana. It was a time before television had invaded the outback, but an enterprising company which was marketing an antenna dish for satellite television installed one at the conference centre, which happened to be the Burke Bowling Club. Excitement was running high at the prospect of viewing the royal wedding.

A big television screen was set up at one end of the Bowling Club bar. Long before the wedding was due to start, all of the women present had taken chairs and planted them firmly in front of the screen. Most of the men stood at a distance around the bar. Some broke rank and went to sit with their wives, accompanied by the good-natured chaffing of the others.

At the other end of the long barroom, were several banks of poker machines. This area was deserted with the exception of a solitary player who had come from the bush in Western Australia and had never seen a poker machine before in his life. He immediately became hooked and the historic event about to be screened at the other end of the bar was a non-event as far as he was concerned.

The ceremony at Westminster Abbey began. Gradually, the majestic dignity of the occasion drew a reverent silence over the room. Even the drinkers at the bar half-turned to watch the proceedings. The ceremony came to the climax where the bride and groom exchange vows. Everyone leaned forward in their chairs as the big screen was filled with the intimacy of the moment.

Suddenly the sacred silence was broken by a cacophony of violence, the ringing of bells and the metallic clatter of showering coins. The man from Western Australia had hit the jackpot! The rest of the room turned and looked at him as if he had committed the blasphemy in Westminster Abbey itself. But the man from Western Australia was oblivious of their anger. Dame Fortune was his bride and he would remain with her for the rest of the night.

No-one could begrudge the people of the outback the right to the times of enjoyment at the conference. More often I found myself confronted with reminders of the stark realities they had to face in their isolation.

'If you happen to be in Adelaide in the near future,' said one man, 'I would be grateful if you would look up a mother who is down there with a very sick child.'

I promised I would try to make contact. A few weeks later I was in Adelaide and went to the address I had been given. To my astonishment it was a caravan park on the outskirts of the city. Most of the vans were hired to people who could not afford to rent a house.

The woman in the caravan had come down from the desert country west of Broken Hill. She had no relatives in Adelaide and no friends. Her daughter had a serious disease, her life was at risk and her hospitalisation was long-term. A rented caravan was all the mother could afford. The husband had remained on the property because there was no-one else to look after it.

I asked the mother how she coped with living by herself in a caravan.

'I hate it,' was her bitter reply. 'They talk about me living in isolation up on the property, but this is a thousand times worse. A half-empty caravan park is no place to come home to when you have just left a dying child.'

The vulnerability of children in the isolation of the outback became a matter of deep concern for me, but there was another group of people in the bush who also tended to be forgotten.

These were the elderly. Some lived quietly with their families on the stations where they themselves had grown up. Others lived alone. I met elderly couples continuing to battle away in the bush. To take them away from their country would kill them. Sometimes I met a lonely old man who, having lost his wife, didn't want to lose his world.

Such an old man lived in the distant reaches of the Northern Territory. His wife had died a year or two before. One of our padres named Keith called in to visit him. Keith and his wife Rhonda travelled around that part of the outback in a specially built campervan. They lived in it for nine months of the year. The first time they visited the old man he was still grieving and very agitated at the thought of entertaining visitors. Keith and Rhonda didn't stay long but promised to call again.

On their next visit, they received a warm welcome and were invited to stay the night. They sat down and talked for a while. Then the old man muttered something about getting dinner ready and disappeared into the kitchen. Rhonda sat and listened to the pots and pans being knocked around. Eventually she got up and went into the kitchen.

'You go and talk to Keith,' she ordered. 'He gets enough of me during the day.'

The old man didn't need much encouragement. When eventually Rhonda brought in the dinner, the old man and Keith were engaged in an animated discussion. They sat down to eat and the padre said grace.

'Well,' said Rhonda, as the old man picked up his knife and fork, 'I hope it's to your liking.'

The old man looked her straight in the eye.

'Lady,' he said, 'for the company, I'd eat it raw.'

12

The Billabong Kids

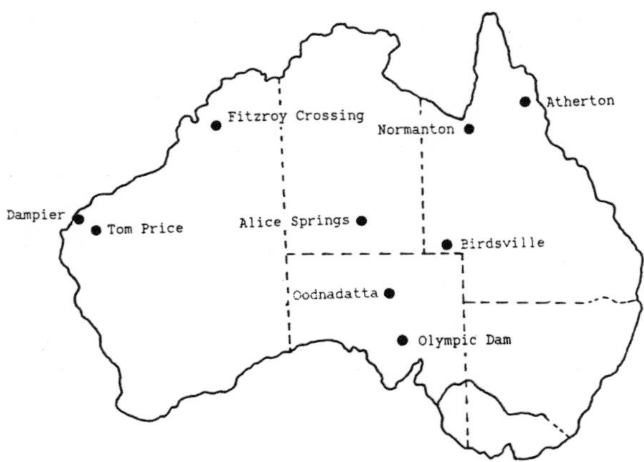

'The Aboriginal camp is over there,' said the station manager, waving his hand in the direction of the bush. 'You can walk over if you want to see it.'

I walked away from the homestead, across the cleared space surrounding it and came to the edge of the scrub. The trees were scraggly and withered. As I walked between them, the ground beneath my feet crumbled like overcooked pastry. When the breeze lifted, so did the dust. The sun was hot and the insidious humidity began to dampen my body.

After a while I came to a large open space where the Aborigines had their camp. I looked around. The ground was dry and dusty. It was not my kind of camping place. The Aborigines sat on the ground in small groups. Camp fires were smouldering, spilling grey ashes on the ground. Black billies simmered around the edges. On one fire, there was a saucepan filled with large pieces of raw meat.

People drifted around with no apparent purpose. Occasionally voices uttered crackles of conversation in that sharp tone which is uniquely Aboriginal. There seemed to be an atmosphere of listlessness and aimlessness.

Then I saw a small Aboriginal boy playing by himself. It was a game I think they call skip stones. He played with that innocence and intensity of childhood that is oblivious of everything else. His eye was uncanny as it is with most Aboriginal children before trachoma, or one of the other common eye diseases, begins to pull down the blinds.

After a while the boy stopped playing and ran over to one of the groups sitting on the ground. He ran with a loose-limbed ease. The group appeared to be his family. He sat next to one old man, who was probably a grandfather or an uncle. They presented a picture of intimacy that often marks the relationship between the very old and the very young. The boy sat for a while and then got up and resumed his playing. It may not have been my idea of a camping place, but it was his place and he was with his people. He was happy.

A few days later, I saw the boy again. This time it was in the little outback town where I was staying. The town was my kind of place, a few pleasant tidy streets lined with shady old gum trees. The buildings were squat and brown. One or two had verandahs stretching over the footpath. Outside them, a few dust-covered bush vehicles were parked. Some of the station people were in town buying stores.

I glanced into one of the shops and saw the Aboriginal boy from the camp. He stood alongside his mother, almost folding himself into her ample cotton frock. His wide eyes were bright, not with happiness, but with the hard gleam of apprehension. There were a few white people in the shop. This was not his place nor his people.

Later in the day, I saw the boy again. I was standing in the clinic of the town's hospital. One of the nurses was treating a small queue of patients. Most of them were Aborigines. The boy came into the clinic room with his mother, where the nurse greeted him cheerfully. He had an ear infection that needed treatment. Obviously he had been in the hospital before and did not appear over-anxious. The nurse inspected his ears and dabbed them with some medication. Then she straightened up and stood back.

'There you are, Jimmy,' she said. 'Now be sure to keep those ears clean.'

The boy ran quickly out of the room into the sunshine. His mother stayed for a few minutes, chatting to the nurse. Then, with a big smile, she went out after her son. They walked along the dusty streets towards the edge of the town. Eventually they came to a number of small houses that were occupied by Aboriginal families. Some of the woman's relatives lived in one of them and she had gone there to wait until the truck came to take her back to the station and the camp. She sat outside in the shade of a tree and talked with the other women. The children played in the security of each other's company.

Late in the afternoon the station truck pulled up outside. The woman and the boy got up unhurriedly and walked out to the street. They climbed onto the back of the truck and it drove off with a belligerent roar that is the trademark of bush vehicles.

The nurse at the hospital told me that Jimmy's father was a stockman. During the mustering season, he was out at the station and the family was with him. After mustering, the family moved back to the town. It was a curious nomadic life.

Jimmy was four and next year would be eligible for school. I wondered how the boy from the bush camp would cope with the classroom. The headmaster at the town school told me.

'It's very difficult,' he said. 'Some of the Aboriginal children who come here have never sat on a chair, let alone at a table. Their parents don't do the things that white parents do in preparing their children for school. The Aboriginal children have never

Good Samaritan of the outback. AIM padre helps restore an overturned vehicle in the Kimberley.

White minority. Children of Halls Preschool Centre, Halls Creek.

One of the Billabong Kids. An Aboriginal boy hurrying to Halls Creek Preschool Centre.

played with coloured pencils or read a picture story book. To come and sit in a room at a desk for hours on end is an enormous adjustment for them to make.'

He paused for a moment and then went on.

'Some of them are very bright. They learn quickly. But it doesn't help when, like Jimmy, they just begin to settle in and then get whisked off to the cattle station for the mustering season. Some of the stations have little schools, but most of the children miss out and never catch up.'

This was one of my first encounters with the children of the bush. Later, I came to learn that the disadvantages they suffered were not limited to Aboriginal children. I began to think of bush children, Aboriginal and white, as the 'billabong kids'.

A billabong, as Australians know, is a backwater that is created when a bend in a river becomes cut off from the mainstream. The water in a billabong becomes muddy and stagnant, because it no longer flows. The bush children of Australia have lived, and continue to live, under the constant threat of being cut off from the mainstream of social and educational development.

Aboriginal children, caught in the firing line between two cultures, are the obvious victims. But I came to learn that white children suffered the same fate. Take Stephen, for instance.

I met Stephen in another part of the outback at another time. Like Jimmy, he lived on a cattle station. But Stephen was white and the station belonged to his dad. I met Stephen in a children's hostel in a far-north Queensland town. He was aged six and he was there with a young woman who was his mother. She was small and strong but marked beyond her years by the life of the outback.

She sat forward on a chair, her body arched with anxiety. She punctured her mouth with the nervous jabs of a cigarette. On the other side of the table sat an older woman, who looked kind and competent. She was in charge of the hostel. They were talking about the boy coming to live at the hostel. Finally, the young mother asked the question that gave expression to her anxiety.

'He'll be all right? He's very young to be leaving home.'

The older woman smiled reassuringly.

'I think you'll find he'll settle in. Most of the children here come from the stations. One or two are nearly as young as Stephen. He'll make friends.'

The woman from the bush leaned forward to stub out her cigarette and then leaned back as if she had relaxed a little.

'What will he need to bring? We're different out in the bush you know. He brushes his teeth of course, but I suppose he'll need pyjamas and a face washer and things.'

The older woman smiled again.

'I'll give you a list of things to bring. But don't worry. We have a good supply here and he won't want. They get a little pocket money every week so it would be helpful if you could send some. I think that's about all.'

The woman from the bush glanced over at Stephen, who was comfortably curled up in an armchair, reading a comic. He seemed unaware of all the concern being expressed about him.

The interview came to an end. The older woman took the younger one on a tour of the hostel. The bedrooms were bright and airy. There was a large dining room with a kitchen at one end. Some of the older children were peeling vegetables for the evening meal. At the other end there were armchairs and bean bags and a television set.

'We don't have television back at the station,' said the woman from the bush.

Soon after, she took her leave. She walked out to where she had parked the truck. Across the road from the hostel was the school Stephen would attend. It was set in wide grounds, but the buildings seemed large and forbidding. She looked back at the hostel and at her son. Then as if she had made a decision, she switched on the engine and drove slowly down the wide street which led to the highway.

On the way back to the station, she had plenty of time to reflect on what had happened. It had been a hard decision, but what else could she have done? Her husband had said they couldn't afford a cook for the stock camp this year, which meant she would have to do it. That meant living in the camp for six weeks.

Of course the boy could come with them. He loved living out in the camp with his dad and the other men. She supposed it would be good for him. But it wasn't what he needed at this stage. He needed to go to school and have the company of other children. He needed to know there was a world outside the station. Later he might decide to come back and work with his father. But he had the right to know what else he could do and make his own decision.

The thought crossed her mind that she had made a decision like that when she had married Bill and come to live up here. A moment of uncertainty wavered her mind, but she quickly brushed it away before it developed into guilt. She began to concentrate on driving into the setting sun of the late afternoon.

Back at the hostel, I walked around with the older woman while she folded sheets and put out fresh towels. I asked her about the woman from the bush and the pain of parting with her six-year-old son.

'I go through this with parents every year,' she answered.

'They have to decide at some stage what they want for their children. Inevitably that means parting with them. Some leave it as late as possible. Others, like that woman, really have no choice. I should be used to it by now.' There was sadness in her voice. 'But somehow the pathos of it all never ceases to affect me. Bush parents have a lot of hardships to overcome, but the hardest is parting with their children.'

A week later, the woman from the bush drove her son to the front gate of the property. She parked it under the shade of a solitary tree. For a while, mother and son sat in the companionable silence of the bush which they knew so well. They were waiting for the truck which went past every week, up the road to the town with the hostel.

'You've got your toothbrush?' she asked.

He didn't answer because she had packed it herself.

The sound of the approaching truck began to emerge. Its growling grew until a cloud of dust appeared on the horizon. Then, all too quickly, it pulled up alongside them. The woman and the boy got out and pulled an old suitcase from the back of their vehicle. The truck driver stowed it in the cargo on the back.

The boy scrambled up to the cabin of the truck. Halfway, he suddenly remembered that he had not kissed his mother goodbye. It was too late now.

'Take care of yourself, Mum,' he shouted down at the anxious lined face looking up at him.

Her reply was lost in the roar of the engine, as the driver swung back on the road. The woman from the bush watched as the truck diminished into the distance. Then she turned and walked back to her vehicle and sat for a moment, bent over the wheel, before straightening up and driving back to the homestead.

The hostel manager gave Stephen some special attention during the first few weeks he was there. They didn't spoil him and he had to take his share of the duties to be done. Not that he minded that. Like most bush children he was used to hard work. It was expected of them on the station and they didn't mind. Everyone worked hard on the station. By the time they were ten or twelve, bush children were treated like adults. In some ways it was hard to come away and be treated like children again.

But often, in the end, it all became too much for some of them. At the beginning of the next term or the next year the boy or the girl from the bush didn't come back to the hostel. There were letters from the parents, explaining that the children were going back on correspondence, or that it had been a bad season and they couldn't afford the fees. The real reasons were much deeper and harder for city people to understand.

The net result, however, was an addition to the number of the billabong kids.

Aborigines and station people are not the only inhabitants of the outback. There is another large group of people whose children are affected by living in isolation. These are the people who work in the mining industry. The outback is dotted with mining projects, some already operating and some in the development stage.

I met another small boy about the age of Jimmy and Stephen. He lived with his family at a site where a vast ore deposit had been discovered. His father was a member of a drilling team which was testing the extent of the deposit and looking for the best place to commence mining.

There was a time when a drilling camp was a very basic, all-male establishment. In recent years there has been a growing tendency for the drillers to take their families with them. They became fed up with being separated by thousands of kilometres. It meant living in fairly primitive conditions, but the families reckoned that is was a small price to pay for being together. They attached their caravans to their trucks, drove to the drilling site and made their own camp. There was generally a single men's camp on site and a mess where everyone could eat.

I drove to this drilling camp with some friends from a neighbouring outback town. The country over which we travelled was endless desert sand with sparsely scattered stunted scrub. Our faces were whipped and burned by sand and sun as we drove. It was early when we arrived at the camp, but the temperature was already rising fast as we pulled up at a cluster of caravans. There were some makeshift clotheslines from which rows of working clothes were flapping vigorously.

Outside one of the vans, a woman and a small boy were waiting for us. The boy's name was Mark. He was five, small for his age and wore a serious expression. We went into the caravan and talked for a while with his mother. The inside of the van was scrupulously clean. The woman suddenly got up and went to the door. She called to the boy.

'Mark, it's time for school.' Then she turned and spoke to us. 'Today is his first day on School of the Air.'

The boy came in and sat down at a table which had been placed at one end of the caravan. There was a small radio transceiver on it. It was already turned on and was emitting the muffled crackle and static of radio reception in the outback.

Then the transceiver came to life. It was the School of the Air teacher greeting his students. The children were scattered across the countryside, invisible to each other as well as to the teacher. He spoke to them as if they were all in the same classroom. One by one, the children answered the roll call. I tried to imagine them, sitting in their solitary places in homesteads or in tiny settlements.

Finally the teacher told the children that there was a new member of their class who was starting today. His name was Mark and he lived in a drilling camp. Mark sat nervously still. Outside, the wind howled and beat its frustration against the caravan's wall. It didn't make listening to the radio any easier.

'Say hello to your classmates, Mark,' said the teacher encouragingly.

The boy sat, clasping the handset, but the words didn't come.

'Mark, are you hearing me?'

Mark's mother came quickly across and stood beside him. He looked up at her anxiously. 'Do I press it now?' he asked, his finger on the button. His mother nodded. He pressed the button and after a nervous pause, spoke.

'Hello.'

Mark had begun his first day at school.

School of the Air is one of the miracles that makes living in the outback a little more tolerable. But it is severely limited by the amount of time one channel can provide for any level of school education, let alone any one child. It amounts to about half an hour each day of transmission for any one 'class' of the kind Mark had just joined.

I looked out the tiny window of the caravan across the empty desert and wondered where Mark went at playtime and lunchtime, when children at school create those wonderful make-believe games. When the session was finished, I asked Mark's mother about his future.

'It's a real worry,' she confided. 'Stan and I believe it's very important for the family to be together. Gillian is nine and soon we must make a decision about her future. I only hope it's not too late.'

Children often provide us with our most vivid memories. In that respect, the billabong kids of the Australian outback are imprinted on my memory. But there was another child in the desert of another country, who first made me conscious of how much they are disadvantaged.

The desert was in Arizona, U.S.A. I was travelling through an Indian reservation and had just visited an Indian school. The principal told me of the high truancy rate among the children. They were not engaged in any mischief. They simply didn't come. The fact that they were missing school was not the same cause for anxiety among Indian parents, as it was for the whites.

I left the school and drove for a while through the desert. The track wound its way among the cactus as if it were a slalom run. At the side of the road in front of me I saw a small flock of sheep, so I slowed down. Sitting on a rise overlooking them was an Indian boy. He sat with a colourful blanket around his shoulders and he wore a large sombrero. I stopped the car and walked over to him. At closer range he looked about twelve.

'Is this your regular job?' I asked. He said it was.

'What about school?' He said he had been, but not now.

My logical mind said the next question should be, 'What do you want to be when you grow up?' but somehow it seemed absurd. It struck me that this is how it was for King David, of Bible fame. He began life as a shepherd boy and ended up as a king. I thought about the possibility of this boy becoming President of the United States of America, but I didn't like his chances.

The same thought struck me a thousand times over when I came to know the billabong kids of the Australian outback. It was at a time when I was trying to discover what was the difference between the outback of the era of Flynn of the Inland and the time in which I was beginning my work. I asked Fred McKay, who succeeded Flynn, what he thought was the difference. He gave the question his usual deliberation and then answered in one word.

'Children.'

The struggle for survival in Flynn's day was so grim that most men simply wouldn't take their womenfolk into the bush, let alone children. There were some appalling stories of children who lived in the bush dying for want of medical care. But as Flynn developed his outpost hospitals, his Flying Doctor Service and the pedal radio network, people developed the confidence to go into the outback and raise children.

McKay himself had pioneered many projects to help the children of the bush. A drover once came upon McKay, out in the middle of nowhere, with eight children. He was making an 800 kilometre trip through the bush so the children could have a holiday by the sea. None of them had ever seen the sea before and probably would never see it again.

But it was the lack of opportunity for education that caused McKay most concern. He had children of his own whom he had to leave behind during his long and frequent trips into the bush. It was an experience John Flynn never knew.

Fred McKay worked with a former Governor-General of Australia, Sir Paul Hasluck, to establish a place in Alice Springs where children from the bush could live and go to school without having to travel thousands of kilometres to the capital cities. The place was called St Philip's College. It started as a residential home away from home for bush children and later became a full teaching college.

The value of St Philip's was revealed to me during a visit to Birdsville. The nurses at the hospital generally had something for me to do. On this occasion, they asked if I could go and visit an Aboriginal woman.

'She's very upset,' they warned.

I wandered along the broad boulevard that is Birdsville's main street and turned down a side track. At the end of it was a small house. Beyond its back fence stretched the Simpson Desert. I went up to the door and knocked. It was opened by an Aboriginal woman. She appeared to have been crying. I explained that the nurses thought I might be able to help.

At the mention of the nurses, her original look of suspicion softened. She opened the door a little wider and I walked in. The room was dark and sparsely furnished. We sat down on two upright chairs at a bare table. The mother called out some command in the shrill Aboriginal tongue. Two Aboriginal girls in their early teens slipped into the room and, without looking at me, went over to the other side and sat on the floor.

The story came out from the mother in short angry bursts. The girls had been attending the little primary school in Birdsville. The teacher was young and enthusiastic. He told the mother that the girls were very bright and should have the opportunity of good secondary education. It would mean going away, but he would be glad to help with the arrangements. With some trepidation and personal sacrifice, the mother allowed them to go south to a well-known boarding school.

'Tell the man what happened,' said the mother to the girls.

They sat with lowered heads and shuffled their feet on the floor. Then in a voice I could hardly hear, one of them said, 'They sent us home.'

It was not easy to get the whole story. I gathered that the girls had been at the school for about one term. Then, without warning, they were called in and told that they were obviously not fitting in and perhaps it would be better all round if they left.

Through some slip-up at the school, the mother was not informed. The first she knew was when the girls flew back on the weekly plane and suddenly appeared at the front door. The effect on the girls was devastating. It was as if someone had opened the door to a bright future and then slammed it on their face. By the time they had finished telling me, everyone was in tears.

I left the house and said I would come back later. I went down to the school and talked to the teacher. Just how good were the girls? Had he built up expectations that were unreal? He assured me that the girls were very intelligent and, given the right encouragement, would do well. I went back to the house and talked with the mother.

'We have a place in Alice Springs,' I said. 'All the children come from the bush and there are a number of Aboriginal children. The girls will feel much more at home.'

The mother laughed.

'You want them to walk over the Simpson Desert?' she asked. 'And then when you throw them out they walk back again. That's what happens to Aboriginal people.'

We talked for a long time. I explained that friends would fly the children across to Alice Springs and bring them back at the end of each term. If Flynn could fly people with critical medical needs, then why not children with critical education needs.

I left the mother to think about it and talk with the girls. Aboriginal people have their own decision-making processes and I was not part of them. The girls went to the college in Alice Springs. Despite the Simpson Desert, The Alice is closer to Birdsville than Brisbane, in more senses than one. The door which had been shut in their faces had been opened again, at least for a while.

Like most parents with small children in the mid-60s, I was one of the fathers who queued up to enrol my sons for kindergarten, or preschool, as it came to be called. Kindergartens were sprouting like mushrooms on the edges of the mazes of metropolitan suburbia. We were solemnly told that these were the most important years in the life of a child and that early intervention would save children from untold agonies in later years. We believed it and the kindergartens were filled.

So it was not surprising that where John Flynn had built outpost hospitals in places like Oodnadatta and Birdsville, Fred McKay, who succeeded him, built kindergartens in the new towns of the outback, fifty years later.

Mining towns like Tom Price and Dampier were filled with young families. Many of them were the kind of people who, had they remained in the cities, would have joined the queues and the waiting lists of the kindergartens. To come into a raw new town in the desert and find a kindergarten being built was a pleasing and gratefully accepted surprise.

Aboriginal parents, on the other hand, had no such understanding or expectation. Yet in many ways, Aboriginal children had greater need of preparation before they were placed on the conveyor belt of Western education. Early intervention in the medical care of Aboriginal children was beginning to have significant success in the Kimberley region. It was the nurses who first recognised that the same kind of help might be appropriate for their education.

It was hard for me not to agree with that proposition. By this time, we had six or seven years' experience running kindergartens in remote mining towns. But I wasn't sure it would be of much help when it came to Aboriginal chidren.

A kindergarten teacher in a Pilbara mining town told me about one of her children, aged four. He approached her with a painting he had done and asked if she would like it. She solemnly accepted the painting and said she would hang it on her office wall. A few minutes later the boy came back again and asked if they could have a private conversation. The teacher closed the door and asked him what he wanted.

'Well, since I gave you my painting, I thought you might give me something.'

'Like what?' asked the teacher, somewhat puzzled.

'Like money.'

'How much?'

'About twenty,' said the boy, without blinking.

Somewhat deflated, the teacher reached for her purse and took out a twenty-cent piece. The boy regarded it with disdain.

'Not cents,' he said. 'Dollars!'

I didn't think we would have to deal with that level of sophistication when it came to dealing with Aboriginal children, at least not for a while.

The kindergartens we built in the Kimberley region for Aboriginal children might at first glance appear little different from those in the cities. They had the same miniature kitchens, a dolls' corner and a place where the teacher sat surrounded by the children while she read them stories. The toilets were child-size and the hand basins closer to the floor. Even some of the songs they sang were reminiscent of those I had heard my children sing.

But there were differences. You noticed how the one or two white children who attended stood out as the colour minority. Two Aboriginal mothers were cutting fruit, but the smell of freshly cooked damper coming from the kitchen was not one which normally permeated a city kindergarten.

There were other differences of a basic kind. One day I walked into the kindergarten at Fitzroy Crossing. There were two pieces of white butcher paper stuck to the wall. One was headed: 'Things to learn this week. Children.' The other was headed: 'Things to learn this week. Teacher.'

At the top of the teacher's list were the words, 'How to kill and cook a dingo.' I looked at the teacher and asked if it was true.

She nodded and laughed. 'It happens this afternoon and I can't get it over too quickly.'

I went along with them to learn. We all climbed into the little kindergarten bus and drove into the bush. The children were excited. They pointed to things in the bush I could not see. The bus weaved its way along the sandy track and brushed the overhanging trees. Eventually we came to a slope that led down to the bank of a river. We stopped the bus and got out. The children ran straight for the river.

There was an Aboriginal man walking slowly through the long grass, dragging something on the ground behind him. It was a dead dingo. We had missed the first part of the lesson. The Aborigine, Johnny Marr, laid the dingo on the ground and the children crowded round. The teachers and mothers were not so keen. Johnny then produced a knife and with great dexterity proceeded to skin and gut the dingo. Not everyone enjoyed the lesson.

The kindergartens served the children who lived in the towns, but there was another group of Aboriginal children whose needs were even more desperate. They lived out on the stations.

Some like Jimmy, the little Aboriginal boy whom I had seen in the hospital, lived in the town when his father was not out at the stock camp. There were many others who never left the station. But not even distance could insulate them from the cultural conflict. Sooner or later the Aboriginal children on the stations would encounter the impact of Western culture.

So, after the first kindergartens were opened in the Kimberley towns, we introduced a mobile preschool service for the children on the cattle stations.

Lorraine was the first teacher to do the station run. She was slightly built and the big four-wheel drive she was given seemed to swamp her. I travelled with her on one of her early runs. There were three cattle stations to visit. The round trip covered 400 kilometres and took three days. The bush tracks were notoriously rough. The night before we left, we packed the van. It had everything from basins to bean bags. I remarked that I was glad I didn't have to load and unload every day.

'Don't speak too soon,' said Lorraine.

She drove the four-wheel truck with competence, but it was hard going. We bounced and bucked our way along the tracks in a way that would have tested endurance rally drivers. Eventually we arrived at the first station. Lorraine drove slowly down the track and stopped in a clearing under a tree. There was no-one in sight. I began to unload the equipment.

'Don't take everything out,' said Lorraine.

Then there was the patter of little feet and a dozen children burst onto the scene. Behind them at a distance came the Aboriginal mothers, walking with a slow, easy gait. The children picked up some of the equipment and began to play. One girl found a black doll and was intrigued. A boy grasped a toy tip-truck. He examined it, placed it on the ground and gave the wheels an experimental spin. He laughed and spun the wheels again. Then he scampered around the ground pushing the truck at full speed.

'It's probably the first time he's ever seen a toy with wheels,' said Lorraine. 'That's why I never put all the equipment out. They get too excited and can't cope.'

The mothers were sitting on the ground. Lorraine went over and told them what she was doing with the children. When it was all over, we repacked the equipment into the vehicle and drove down to the homestead. The manager and his wife were hospitable and seemed happy to see Lorraine come and teach the children. But I felt the manager still had reservations about all the money being spent on the Aborigines.

Next morning we drove out and on to the next station. The tracks if anything were worse. By this time the dust had a firm grip on the interior of the cabin. All the equipment we unloaded appeared to be heavier, or maybe my enthusiasm was diminished.

There were about twenty-five children at this station and they seemed to be of all ages. I asked Lorraine how she coped with such a diversity of children. She smiled wearily.

'It's very unprofessional,' she replied, 'and if one of the supervisors from the city saw me teaching children of that age . . .' She looked at a boy of about fifteen. 'I'd be dismissed on the spot.'

'But you know,' she continued, casting another look at the shy fifteen-year-old who was standing back from the others, 'I can't turn them away.'

The trip to the third station was the longest and hardest. It was also a waste of time.

'They're not here,' said the station manager, after we had searched fruitlessly for the children. 'They've gone into the desert.'

He didn't invite us in for a drink.

'He doesn't like Aborigines and he doesn't like women,' said Lorraine as we drove away. 'But I keep coming back.'

The return trip was long and the day was hot. The dust rose up from the churned track and found its way into the cabin. We agreed it would be nice to get back and have a cold drink and a shower.

'You'll be glad to rest up after a trip like this,' I commented sympathetically to Lorraine.

She laughed.

'By the time I've unloaded the equipment and cleaned it, hosed down the vehicle and done my own pile of filthy clothes, all I want to do is collapse.'

I offered to hose down the vehicle and clean the equipment. My sympathies were with the teacher, but my thoughts were with a fifteen-year-old boy who didn't know what it was like to go to school.

There are white children for whom the isolation of the outback can be just as crippling. Take Ben, for instance. He lived on a station in the far outback of Western Australia. The station had very few visitors. One was the local padre who visited occasionally with his wife.

Ben's mother became very ill and had to be taken to Perth for an operation. His father was at his wits' end. He had no help on the station and simply had no time to look after Ben. The padre and his wife offered to take the boy until the mother was well again.

I happened to be visiting them a few days after Ben came to stay. He was nowhere in sight when I arrived and didn't appear for dinner. I remarked that I didn't usually have this effect on children.

'Don't worry,' said the padre's wife. 'He simply isn't used to people. And it's all a bit much for him. His mother rang up from Perth last night to speak to him. Ben's never seen a telephone before. He heard his mother's voice and looked around the room wondering where she was. Then he just burst out crying.'

The Bens of this world grow up, but the isolation holds them down. I was visiting an outpost hospital in the far west of Queensland. The nurses asked me to conduct a church service. It was something the padres usually did, but I was happy to take my turn. The nurses made contact with all the stations in the area and let them know the service was on.

I stood on the long verandah of the hospital and met the people as they arrived. They dismounted from dustry trucks and self-consciously brushed down their clothes.

A truck arrived which looked as if it had come in from the desert. It was caked hard with mud and the windscreen-wipers had given up the unequal struggle. A man got out of the truck. With him were three young people. The boy looked about sixteen and the two girls were eleven and eight. They stood uncertainly for a moment and then walked slowly towards the hospital.

'They're living out in the desert,' said one of the nurses. 'He's a fencing contractor and the job will keep them there for six months.'

The man introduced himself and shook hands. He apologised that his wife was unable to come. 'She's not well and the trip in is a bit rough.'

After the service people stayed for afternoon tea and I talked with the children. I asked the fifteen-year-old boy how he got on for school. He looked a bit embarrassed and said that he helped Dad with the work.

'We're pretty busy,' he said, 'and I just had to stop doing the correspondence lessons. Anyway, we only come into town every couple of weeks and then sometimes miss the post.'

After they had gone, I asked the nurses to tell me more about the family.

'They all work out there,' said one of them. 'The eight-year-old girl can drive a big truck as well as a man. The mother tries to keep them up with correspondence lessons, but she hasn't been well.'

As they talked on, it seemed that the nurses were torn between an admiration for the close ties which kept the family together and a deep concern for the children's future.

'Sometimes I think it's such a waste,' said one of the nurses.

'How do you think the parents feel?' I asked.

There was a long silence.

'I think they just can't bear to let them go,' was the reply.

Even when parents eventually let their children go, it can be too late. Kevin was fifteen when he first left home to go to school. He had lived with his parents on a property which was hundreds of kilometres from the nearest school. In the wet season, the homestead was cut off from the rest of the world for months. This made correspondence lessons almost impossible.

The homestead was located in a place where radio reception was poor and School of the Air difficult to sustain. Nevertheless, the family struggled on with Kevin's education.

By the time he turned fourteen, his parents realised they had to do something. I met them during a visit to the region and they told me of their concern. I encouraged them to send him to a hostel which had just been opened in a town which had a good high school. In North Queensland terms, it was not all that far away from their home. They decided to send him.

The next time I visited the hostel I asked the couple in charge how Kevin was faring. They told me the story.

The high school had a uniform for the children. The boys wore grey shorts and shirts. It was a sensible dress for the climate, but back on the station Kevin's everyday dress was his working clothes, jeans, check shirt, riding boots and a stockman's hat. On his first day at school he wore his working clothes. The teacher was polite but firm. Kevin had to go back to the hostel and change into uniform. Some of the children in the class giggled. Kevin stormed out, red-faced. As far as he was concerned, it was his first and last day at school.

The hostel parents talked to the school principal and commonsense prevailed. He was allowed to come to school in his working clothes. The hope was that when he felt more at home he would wear the uniform of his own accord.

Kevin went back to school, but his stay was short-lived. The adjustment from being a man on the station to a child at school was too much. He went back to join the ranks of the billabong kids.

Statistics show that the percentage of children in the cities, who start secondary education and finish it, is about 80 per cent. The percentage of bush children who start secondary education and finish it is about 12 per cent. The story behind those statistics is the story of the billabong kids.

13

The Old-Timers

He leaned forward from the log on which he was sitting and stirred the smouldering coals under the blackened billy.

'People often ask me what was the most important thing that ever happened to the outback,' he said slowly. He leaned forward again and gave the coals another poke. 'And I always say to them, the most important thing to happen to the outback was John Flynn; Flynn of the Inland.'

We sat in silence for a while, watching the thin stream of blue smoke curl up into the air. The shade from the overhanging gum tree made blotted patches on the yellow earth around us. Overhead, the blue sky reigned supreme.

'The second most important thing to happen to the outback was the kerosene refrigerator, and the third thing,' the old man sitting opposite me concluded with a grin, 'was Sunshine Full Cream Powdered Milk.'

Bob Gregory was a bushman of Central Australia. He had been born there and would die there. We had gone out into the bush to boil the billy and have a bit of a yarn; 'to get away from the women for a while,' as he put it.

Bob's home was on the outskirts of Alice Springs. He had been talking about the man who gave him that home; the man they called Flynn of the Inland; the man who was 'the most important thing to happen to the outback'.

Bob's home was called Old Timers. He and his wife lived in an old cottage, set in the grounds of a twenty-six-acre property. There were about twenty such cottages, some old and some new, spread around under the shady old gum trees. In the centre of the property, surrounded by stretches of green lawn, was a long low building with a central courtyard. It was a fifty-bed nursing home.

Old Timers had long been a dream of John Flynn. Thanks to his work in establishing the outpost hospitals and the flying doctor service, the tragedies which took the lives of many inlanders diminished. More people survived and went on to live to a ripe old age. It was for them that Flynn and his mates built this place called Old Timers.

The flying doctor, the nursing hospitals and the patrol padres were Flynn's contribution to the battle for survival in the bush. Old Timers was a symbol of that survival. It was also Bob Gregory's home, and mine too, or at least my home in the inland.

I was a long way from collecting the old age pension. But I visited Alice Springs often and Old Timers was the place where I stayed. I loved yarning with the residents. They were my link with the past, as I went to walk in the footsteps of Flynn.

Flynn had known these old-timers in the early days of their struggles on pastoral stations and in the mining camps. He knew that one day they would have to break camp and leave. But he also knew that it would kill them if they had to pack up and leave the bush altogether.

There was an old-timer called Dick who lived in the back country, beyond Birdsville. Old Dick lived by himself. He would have it no other way. A padre called Les patrolled the region around Birdsville for years and used to call on Dick whenever he was in the neighbourhood. Les built a couple of little cottages next to the hospital in Birdsville. His aim was to give the old people of the bush a place to live when the isolation got too much for them.

So one day when Les had finished building the huts, he went out into the bush in search of Dick.

'Dick,' he said, when he had found him, 'you'll have to give up living here soon you know. You might get sick and there is no-one to look after you. I've built a couple of little places back in Birdsville. You'd have your own bunk and stove and washing trough. It's even got its own dunny.'

Les paused to let that sink in.

'Then of course the hospital's right alongside and if you got crook, the nurses would be there to look after you. Oh, and the store's just across the road for your tobacco.

'Now come on. What do you say?'

Old Dick gazed out into the desert, his lined face and tired eyes still defying mother nature to get the better of him. He thought for a while and then turned to Les.

'Nah, Les,' he answered. 'I couldn't go and live in Birdsville. City life's no good for me.'

It was the fear of leaving the bush that created so much heartache for the men and women who lived there. They knew that the time would come when they could no longer battle on. That was the time when they would hand over the property to their children. The mining prospectors, too, knew that the constant movement around the bush from one camp to another had to come to an end.

John Flynn talked about the need to give the old people of the bush, 'a decent bit of shade'. So he and his mates built Old Timers on the outskirts of Alice Springs.

To find Old Timers, you take the south road out of town and head towards the Macdonnell Ranges. The road runs through a cleavage in the ranges that looks as if it had been cut by a gigantic axe. It's called Heavitree Gap and, despite its narrowness, the road and the railway line and the Todd River all run through it.

On the other side of the gap, further down the road, you can see an upstart of a mountain, standing all by its arrogant self. It's called Mount Blatherskite. The land on which Flynn and his mates built Old Timers lies between the gap and Mount Blatherskite. The Stuart Highway runs along one boundary. On the other side is the dry sandy bed of the Todd River.

The year was 1965 and I was visiting Old Timers, Alice Springs, for the first time. It didn't look much of a place then. There were a few little cottages made from concrete blocks stuck down at one end of the property. In the middle was a longer building, faced with rendered concrete. It served as a hostel.

The matron looked at me disapprovingly.

'Tommy put a tie on especially for you,' she said, 'and here you are without one.'

A cheerful efficient woman, the matron had arranged an afternoon tea party for me to meet the residents. I had made the mistake of believing that the people of the bush didn't dress up for such occasions, but Tommy put me to shame. I went up and thanked him for going to the trouble. That seemed to please him. Sitting alongside Tommy was another old man.

'He was one of the Afghan camel drivers,' explained the matron. 'He's punched his camel teams up and down the track to Oodnadatta more times than he can remember.'

There were about eighteen men and women living in the hostel. Visiting the place was like walking into a living history book. Some of the residents remembered John Flynn and the struggles he had to undergo to build the first hospital in Alice Springs.

'It took him about ten years to get it started,' said one of the old-timers. 'There was one fella, dead against it. He reckoned it was a waste of money and anyway, no nurse was going to lay a hand on him.'

He chuckled reflectively.

'Anyway, Flynn got the hospital opened. The first patient to use it was this fella who had done all the complaining. He fell off his horse and broke his leg. By the time they got him to hospital, he didn't care who laid a hand on him.'

I asked the residents if any of them remembered Bruce Plowman, who was the first padre to visit Alice Springs back in 1915.

'Yes, I remember the young man,' said one old lady, slowly. 'He came up from Oodnadatta with a team of camels. It was a long slow trip in those days.

'Mr Plowman caused me a great embarrassment,' she went on, obviously enjoying telling the story. 'He arrived when I was up on the roof of my house, fixing a hole. He stood there talking to me until I said to him, "Young man, go away and let me get down decently and then I'll give you a cup of tea".'

'How did the place come to be called Old Timers?' I asked.

'It all began long before this place was ever thought of,' said one old-timer.

I reached for a chair and sat down to hear the story.

'It was in the '30s. The hospital that Flynn built had been going for a few years by then. It was right in the centre of the town and everyone used to drop in there. A lot of people who live in Alice Springs were born there you know.'

He paused for a bit as if collecting his thoughts.

'It was about a month before Christmas. A bush fella had been in the hospital one time when he was sick, and he always dropped in for a chat when he was in town. He generally brought a little something for the sisters. They lived on the smell of an oil rag you know,' he added, looking at me as if I were to blame. 'This time, the bush fella went straight up to one of the sisters.'

' "Sister," he said. "I've been thinking." Well, the sister looked at him carefully. Bushies have plenty of time to think, so when they have something to say you can reckon it's worth hearing. "It's coming up Christmas time," continued the bush fella, "and there'll be parties for the kids. But nobody remembers the old people. I'll tell you what. If I fork out the money and you promise not to tell anybody, will you organise a Christmas party for the old people?" '

The old-timer who was telling me the story paused as he seemed to remember the time vividly.

'Well, the party was held at the hospital and it was a big success. Next year, the bush fella came around again and gave the sisters the money to have another one. It became an annual event. They started to call it, the Old Timers Christmas Party. I think it was one of the sisters who gave it the name.

'Then the trouble began. The party was so popular that everyone wanted to come. The sisters had to decide who was a genuine old-timer and who wasn't. Finally they hit on the idea that to come to the Old Timers party, you had to have lived in Alice Springs before the railway line from Oodnadatta was completed and opened.'

He paused again and chuckled.

'Even that didn't stop the arguments. People who didn't get an invite came up to the hospital and hotly disputed their right to come. But that's the way things work in the bush. Anything for an argument.'

One of the old men offered to take me to see his cottage. We walked down to the far end of the property, where his little cottage stood close to the boundary fence. In the late afternoon, the shade of Mount Blatherskite shielded the cottage from the still-fierce sun. It was a simple building made of concrete blocks. An iron roof extended at one end to provide a covered outdoor space. There was an old-fashioned trough attached to the outside wall.

'That's where I do me washing,' said the old man.

Inside were two rooms. One of them had a wood-fire stove and a table and chairs. The other was the bedroom. The floors in both rooms were concrete.

'The matron stuck up the curtains,' he said. 'Adds a bit of colour.'

I looked around and then we went outside. There were a couple of chairs under the shelter of the extended roof.

'I sit out here most of the time,' said the old man. 'The stove's good in the winter, when I have to go inside.'

We sat down on the chairs and yarned for a bit. The big gum trees stood impassively as guardians of the peace and the blue skies kept the door open to the future. I said goodbye and walked back to the hostel. Halfway, I turned and looked back at the old man, still sitting in his chair and gazing across the Macdonnell Ranges. Half his luck!

In the years after I first visited Old Timers in 1965, things in the outback began to change rapidly. The age of kerosene refrigerators and Sunshine Full Cream Powdered Milk was disappearing into the history books and, with them, the Afghan camel drivers and stockmen who worked with Kidman on the Birdsville Track.

Old Timers was changing too. Electricity replaced kerosene lamps in the old cottages. We built a lot of new cottages which had air-conditioning and solar heating. The bushmen no longer had to track down to the dunny on the boundary fence.

'You wait till the winter,' warned the old-timers, as they watched the glass panels of the solar heating units being installed and shook their heads. They were right. The sub-zero temperatures of an Alice Springs winter cracked some of the glass panels in half.

The changes to make the lives of the residents more comfortable did not change the character of the place. The residents retained their individuality and independence. You can't institutionalise a community which includes a man who had been born in Oodnadatta at the turn of the century and lived all his life in the Centre. Or a woman who had once been a Tivoli ballet girl in Sydney and then came to live in the outback as if she had been there all her life.

I came to look forward to my visits to Old Timers. I would dump my bags in an empty room and walk out onto the long back verandah of the nursing home, where I knew I would find Tommy. The sun slanting down between the overhang of the verandah and a big tree in the courtyard caught the big happy smile that creased the corners of his eyes and never seemed to leave his face.

Tommy's great passion was plaiting stockwhips. He tied the pieces of leather to a corner post of the verandah and leaned back a little as he strained the plaiting. Every day would find him in the same place, working away slowly and methodically until it was time to go in for lunch and an afternoon nap. Tommy was always good for a yarn and delighted in telling me stories of the old days when he was a stockman with Kidman.

'There was this young boy working with us, Aborigine he was. The head stockman was always hollering at him; he couldn't do this right and he couldn't do that right. He couldn't even count up to a hundred.'

Tommy took off his old stockman's hat and scratched his head.

' "Well," the head stockman asked the boy to count the cattle in the yard. "How many we got Jimmy?" he asked the boy when he had finished. "Ninety nine," replied Jimmy. "We got ninety nine."

'The stockman knew there was a hundred in the yard so he asked Jimmy to count again. The same answer came up again. "Ninety nine." He did it a third time and got the same answer. "Ninety nine." "You damned well know there's a hundred in the yard," roared the head stockman. "Why are you always telling me there's only ninety nine?"

'Its beauty and its terror.' The Australian outback combines a rugged beauty with fierce temperatures, devastating droughts and flooding rains.

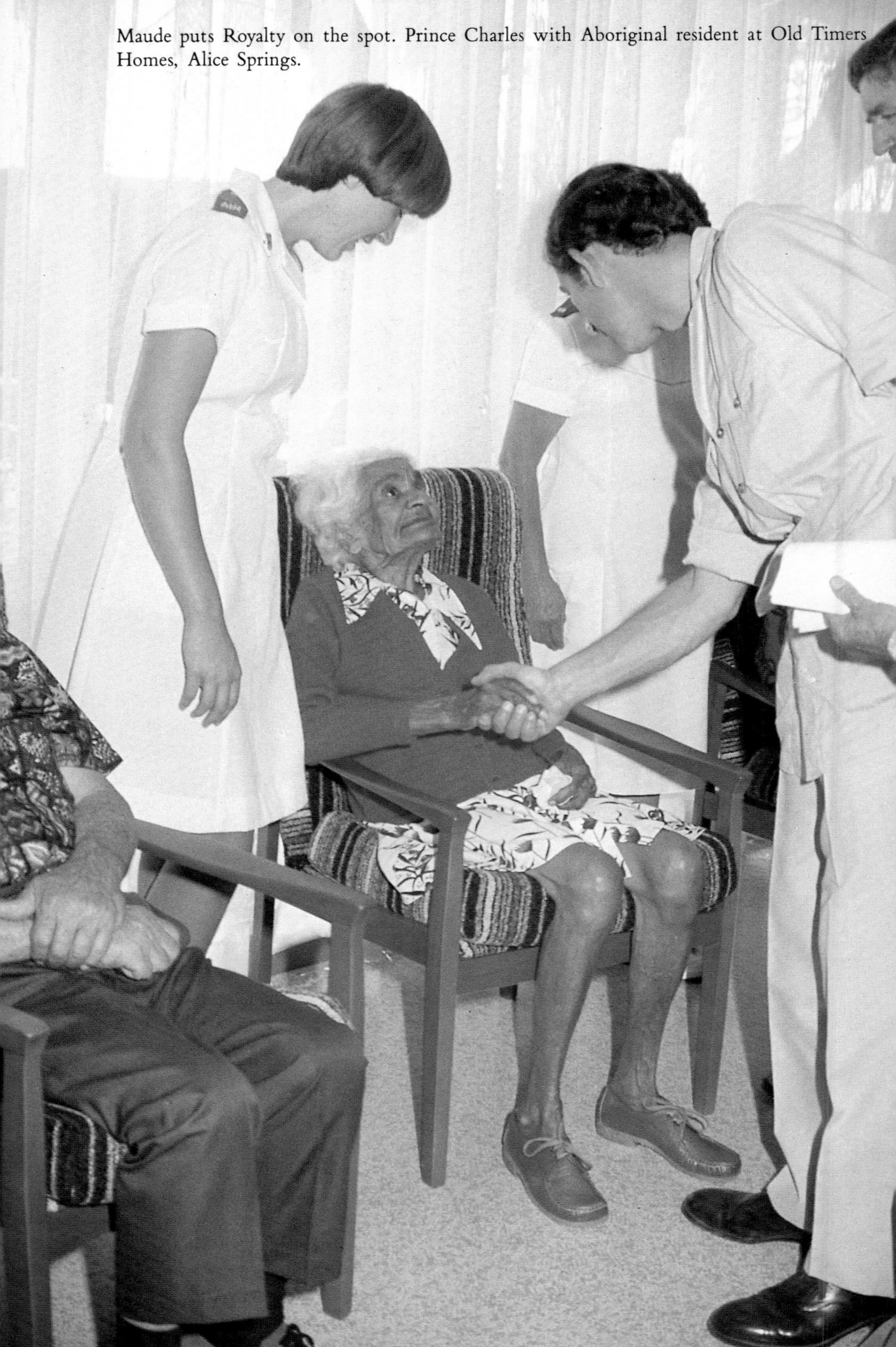

Maude puts Royalty on the spot. Prince Charles with Aboriginal resident at Old Timers Homes, Alice Springs.

' "Well, boss," replied Jimmy with a grin. "You always telling me I can't count up to a hundred!" '

I left Tommy happily enjoying his memories and walked out into the sunshine, passing under the big gum trees that arched across the sky. Down in another corner of the property I saw the slight frail figure of a man. He was wearing the trousers and waistcoat of an old suit and the kind of hat businessmen wore in the thirties. He was leaning on a long-handled shovel, having a breather.

Dick was quite old and suffered from fearful asthma. For a man who had spent most of his life in the outback, he was very pale and looked as if a puff of wind would blow him over. But every morning he came down to this corner of the property and worked on his vegetable garden. Every evening he was down again watering it. I watched him as he dug slowly and stopped often to catch his breath. When I walked up to him, he was leaning on his shovel again.

There was no need to express spurious admiration for the vegetables he grew. The beans were climbing strongly, the tomatoes were plump and red and the lettuce crisp and green.

In the distance, the Macdonnell Ranges ran along the horizon like a long rusty razor blade. Behind us was the inconspicuous hostel, where Dick had a room. Whether in his room or out here with his garden, he could look up and see the ranges, the trees and the sky.

I watched him as he gave the soil a final pat and began to walk slowly back to the hostel. I closed my eyes and tried to imagine Dick in a block of units for old people in Sydney. It didn't fit.

Old Timers had its share of Aboriginal people. Some of them had been there for a long time. Maude was one of them. Maude knew her rights and let the world know. It didn't matter who was around, even if, as it happened, the visitor was Royalty.

Prince Charles once paid a visit to Alice Springs which he won't forget in a hurry. One of the compelling reasons was that he suffered food poisoning after attending a lunch. The other reason was that His Royal Highness paid a visit to Old Timers. He insisted on meeting all the residents and was very charming to them. In due course he came to Maude.

She made a striking picture, with her white hair in contrast to her dark skin. She sat in a wheelchair wearing a red cardigan and had a bright coloured quilt across her knees. I introduced His Royal Highness to her. He bent over, shook her hand and began to speak to her.

'Do you like living at Old Timers, Maude?' he enquired.

Maude looked up at him with unflinching blue eyes.

'Yes,' she said in a loud clear voice.

'How about the food?' continued the Prince, 'Do they feed you well?'

Maude gave him one of her toothless grins. Since she tended to be a bit unpredictable, we were relieved when the Prince began to straighten up and say goodbye.

Then Maude looked him straight in the eye. 'Where's me stubby?' she demanded in a loud voice.

The Prince blinked in astonishment and was speechless.

'Where's me stubby?' said Maude, more aggressively this time.

The bewildered Prince turned to me for help.

'What,' he asked, 'is a stubby?'

'Struth, she's blown it!' I blurted out.

Maude enjoyed a beer. Because she couldn't get out, the staff provided her with a stubby every lunch time. However, she often became impatient and it was not uncommon to hear Maude demanding her stubby for some time before lunch. As the visit of the Prince was scheduled for late morning, the matron had a long talk with her about it. The rest of the staff were more pragmatic in their approach. They gave Maude a couple of stubbies before lunch. It was to no avail.

I explained all this to Prince Charles, who commented that he was sorry he didn't have the time to share one with her.

'I'm going down to cut Tiger's hair,' said the matron.

So I went with her. We walked down to the far end of the property, past the old cottages and climbed through the boundary fence. In front of us, at the foot of Mount Blatherskite, were a few simple houses. This was an Aboriginal fringe camp.

According to the authorities, it shouldn't be there. Fringe camps were considered an eyesore as well as being illegal. The Blatherskite Camp, as this one was known, was a particular embarrassment to the authorities because it stood alongside the main road into Alice Springs and did not give the tourists a good first impression.

We walked towards one of the houses and I noticed two elderly Aborigines sitting outside on the ground.

'That's Tiger,' said the matron, pointing to the old man. 'And the woman is his wife Lily. Tiger, by the way, is blind. I promised Lily I would come over and cut his hair.'

I sat and watched as Gisele, the matron, cut Tiger's hair with great efficiency and chatted to Lily.

'Now don't you forget,' she said to Lily, as she completed her task and we rose to go. 'It's going to get cold soon. So you come over and stay with us.'

'The bureaucrats in Canberra would have a fit if they knew what we were doing,' said Gisele, as we walked back to Old Timers. 'By rights, Tiger and Lily should be over with us, full-time. But it's so much better for them to be with their people. So when the weather's good they stay in the camp. But when it gets cold and wet, or if they are sick, they come in and stay with us. We keep the room at the end of the nursing home for them, so the people in the camp can pop over and visit.'

Someone else climbed through the boundary fence at Old Timers, to visit the fringe camp. But he wasn't about to engage in hairdressing.

It was the tall imposing figure of Malcolm Fraser, the then Prime Minister of Australia. He happened to be in Alice Springs and had agreed to perform the official opening of a new fifty bed nursing home at Old Timers.

It was Anzac Day, so his first official function was to attend the service on Anzac Hill, which stands at one end of Alice Springs and provides sweeping views. As with

Anzac services all over Australia, it was a formal occasion, so when the Prime Minister arrived at Old Timers, he was wearing a lounge suit. Before going inside the new nursing home for the opening ceremony, we took him on a tour of the grounds.

'I've been hearing a lot about fringe camps and the trouble they are causing,' Mr Fraser said as we strolled around. 'I've never seen one. Are there any around here?'

I pointed to the boundary fence at the foot of Mount Blatherskite and said, 'There's one just across the other side of the fence, if you'd like to go and see it.'

Fraser wheeled around and strode off in that direction. His minders, temporarily put off-balance, began making frantic noises into their walkie-talkies. We reached the boundary fence, climbed through and walked over to the camp. There was no-one there.

'Where are the people?' asked the Prime Minister.

'Probably down on the Todd River bed,' I replied. 'It's a nice sunny day and they go down there a lot.'

'How far is that?'

'A few hundred metres or so. It depends how far they have walked.'

I had the sinking feeling that the official opening of the nursing home was not going to start on time, but we set off again and pushed our way through the long grass until we reached the dry bed of the Todd. I looked in both directions and again there was no-one in sight.

'Which way?' said Fraser.

'Downstream I think,' I hoped I was guessing correctly.

We crunched our way along the sandy bed of the river. Then in the distance, I saw a group of Aborigines sitting in the shade of the great old gum trees that line the course of the river. We came up to them. I recognised the Aborigine who was the leader of the group.

'Bluey,' I said to him, 'this is the Prime Minister of Australia. He wants to talk to you about your camp.'

'Sit down,' said Bluey.

We placed our tailored backsides in the sand and Fraser and Bluey talked at length about the fringe camp. Bluey's statements, made in that clipped muttered tone Aborigines often use, were calm and convincing. He said that the fringe camps were important to the Aborigines and they were permanent and not makeshift.

Fraser listened carefully and said he would talk to the Northern Territory government about it. We all stood up and Bluey walked part of the way back with us. Some months later, Fraser said to me that he felt Aboriginal affairs would advance more quickly if negotiations were held on the bed of the Todd River instead of the air-conditioned committee rooms of Canberra.

As we reached the boundary fence, Fraser turned and extended his hand to Bluey. 'Well, goodbye,' he said.

Bluey took his hand and shook it. 'What did you say your name was?' he asked.

'Oh, Fraser, Malcolm Fraser,' replied the Prime Minister.

Bluey nodded and then turned and headed back to the Todd. We returned to the nursing home, where a patient group of people were waiting for the Prime Minister to perform the official opening.

You couldn't always rely on the Todd River to provide a pleasant backdrop for life in Alice Springs.

'The Todd's running!' was a cry sometimes heard in the Alice. It meant that rainfalls further north had come down the various water courses and had caused the normally dry Todd River to run. It was a sight that pleased the locals and fascinated the tourists. Sometimes it was more serious. There were a number of buildings close to the river and some locals occasionally muttered that one day the Todd would rise up and flood them. But no-one really believed it.

Then one day in the early '80s the rains came to Alice Springs with an intensity usually associated with the tropics further north. People shrugged it off and complained with some humour about having to buy mud scrapers for the front door. But the rains continued to come and the intensity, if anything, increased.

My Alice Springs 'mum' was a lady called Nancy. Her husband, George, had been manager of Old Timers for many years. When he died Nancy stayed on and lived in one of the cottages.

Nancy had invited a few friends to dinner and I was one of them. We gossiped about people and the benefit of the rain to the surrounding cattle stations. We opened a bottle of locally grown red wine and discussed its merits with the bravado of people who know little about the subject. Then the lights went out.

Nancy was a quietly resourceful woman. In no time, she had a few candles on the table.

'How will they get on in the nursing home?' I asked the matron, who was with us.

'They'll cope. This happens from time to time. It won't last long.'

We had lapsed back into the comfortableness of chatting about familiar things, when, suddenly, one of the other guests said, 'What is that noise?'

We stopped talking and listened. There was a kind of rumbling noise outside.

'It's probably the Ghan,' someone suggested, since the railway line from the south was not far away.

'No, it's closer than that,' came a reply.

I borrowed a torch, went to the back door and opened it. I shone the torch into the black of the night. Its beam touched the branches of the trees, which glistened with the garment of the raindrops. Then I lowered the beam to ground level. The beam bounced on the rippled waves of fast-moving water. I swung the torch to the left and right. There was water everywhere. The rumbling we heard was the sound of the Todd River, which had burst its banks and was flowing with menacing rapidity. I switched off the torch and went back inside.

'Ring up the police,' said Nancy. 'They'll know what's happening.'

It took a while before the line to the police station was clear. 'It's not good,' said the policeman, when I finally got through. 'The Todd's rising very fast. It's already cut the road at Heavitree Gap. There's no way we can get through to you.'

'But we've got a hundred elderly people here,' I protested. 'And fifty of them are bed patients.'

'I'm sorry. But you'll have to do the best you can.'

'What's the forecast?'

'They say there's a wall of water coming down from north.' There was more than a touch of concern in his voice. 'It's dark and at this stage we have no idea how far the water has spread. We won't know how bad things are until it hits us.'

I wished him luck and rang off. The matron and I took the torch and set off for the nursing home. The water was still rising.

'I'm worried about the people in the cottages,' she said. 'The ground is lower down there.' Then she had another thought. 'Some of the people in the cottages take sleeping pills. And they lock their doors. We might have some problems getting them out.'

The rain continued to pour down as we reached the nursing home. Inside was dry and peaceful. The staff on duty seemed composed, but were anxious to know what was happening. All their patients, fortunately, were asleep, oblivious of the drama that was going on around them.

I went back outside and set off for the nearest cottages. I banged on the doors and called out to the residents to get up and go to the nursing home. The response of the old-timers was varied to say the least.

'Lot of bloody fuss over nothing,' said one old bushie, irritated at being woken. 'It's never come up to here before and it won't this time.'

'Are you coming up to the nursing home?' I asked anxiously.

'No!' he replied abruptly, slamming the door and heading back to bed.

I stood uncertainly for a while, wondering if I should try again. We could always come back and get him if things got worse. But I hoped the others would not prove as difficult. Fortunately they didn't. There was no panic. Some said they wondered what all the fuss was about. But they came quietly.

I knocked on one door and there was no answer. I tried again, banging and shouting for all I was worth. There was still no answer. I judged this to be one of the sleeping pill brigade and added her to the 'see later' file.

Then I headed for the last group of cottages down by the boundary fence. The water became deeper. I reached the cottages and banged on the doors. The residents had already left, so I waded back to the nursing home. By this time the rain was easing and, although it was impossible to see in the dark, I had the feeling that the water had stopped rising. The nursing home patients were still asleep. The cottage people, who had gathered there, seemed quite unperturbed.

'When can we go back to our cottages?' demanded one of them.

'We'll give it another hour,' I replied, remembering that wall of water to the north of the town and the fact that if anything happened, we were on our own.

'Why don't we try Yirrara?' suggested the Matron. 'They're further inland and should be safe. Some of the staff might come over and help.'

Yirrara was a residential college for Aboriginal children. I went outside again, climbed into a vehicle and set out to drive there. It was an eerie feeling driving through the pouring rain, not knowing where the water was flowing and how deep it was. One thing about a bushfire in the night, you can see it.

I reached Yirrara to discover that some of the staff had already left to go to Old Timers. I must have passed them in the dark, so I turned round and went back again. By now, the rain had eased considerably. The water did not appear to have risen any further.

'It's probably safe to go back to your cottages now,' I said to the residents, 'but if you want to stay here, we will fix up beds.'

They left without exception.

Next morning the extent of the flooding could be seen. The Todd had broken its banks and had flooded most of the surrounding country. Water had flooded a number of cottages and caused some damage. The electricity was still cut off, but we were able to get into town and acquire an emergency power plant.

It began to rain again, but the general feeling was that the worst was over. Gisele felt that they had things in hand, and as I had urgent business elsewhere she drove me out to the airport and waved me goodbye. I had not been back in Sydney long before a message came through that Alice Springs was once again being deluged with rain. I made contact quickly.

'It's okay. We evacuated the whole place,' the matron reported triumphantly. 'The whole hundred of them. When it started to rain again, we decided it was time to find out just how difficult it was to get a hundred elderly people out of this place in a hurry.'

'Where did you take them?'

'Over to Yirrara. The Aboriginal children were terrific. They all double bunked and we were able to find a bed for everyone. It was great to see the children and the old folk all cuddled up together.'

'What did the old people think of it?'

'They had a ball,' said Gisele. 'They reckon it's the best outing they ever had!'

The unwillingness of the old-timers to get rattled by the flood reminded me that they had probably faced similar or worse dangers a thousand times.

Old Bob Pasternack was an example. He was originally from Czechoslovakia and had fought in some of Europe's bloodiest wars. He had bullet holes all over his body to prove it. Bob had come out to Australia and gone to work in some of the remote mining operations. He eventually drifted into Alice Springs, as the bushies often do, and finished up living at Old Timers.

Bob was the honorary postie. Every day he put his postie's cap on his head and with a wide smile on his face, set out to do the rounds of the cottages, slipping letters into their boxes and giving a sharp blow on his whistle.

One of his customers was Stan Brown. This old man, now well over eighty, would generally be found on his hands and knees, furiously digging the weeds out of his lawn.

What you couldn't see was that Stan had a wooden leg. He lost his own when still a relatively young man, working out in the bush. It got caught in a pump jack. Stan had to pull it out, knowing that he would lose the bottom half. With the help of an Aboriginal boy, he managed to get on his horse and ride back to the homestead. Then there was a further agonising trip of nearly 100 kilometres to get to the hospital at Alice Springs.

'I'm lucky to be alive,' said Stan, when I complimented him on his courage. 'And you can thank John Flynn for building that hospital.'

Stan owned an old Mercedes Benz. It was his pride and joy and he looked after it with great care. When Mercedes Benz International gave a generous donation to develop a museum in the old Alice Springs hospital and two of the directors from Germany came up to Alice to make the presentation, I asked Stan if we could borrow the old Mercedes to drive the visitors around.

'I'll have it ready,' promised Stan. He proceeded to buff and polish the car to a showroom condition. The visitors from Germany were very impressed. Late in the afternoon, after the ceremony, I drove them back to the airport and waved them goodbye.

It was getting dark as I drove the Mercedes back to town, so I went to turn on the lights. Nothing happened. I tried again. Still no result. Distressed that something might have happened to Stan's car after all his generosity, I took it back and explained that the lights were not working. Stan was interested but not concerned.

'I haven't turned them on for ten years,' was his comment.

I watched the old man park his car under its shelter and return to watering his garden. A short distance away, his neighbour, Bob Gregory, was coming out of his cottage door with a saw in his hand and a determined look in his eye. Further away, in the garden of another cottage, a tall slender man stood among a lush collection of bushes and plants. John Blakeman was the Northern Territory's leading authority on its flora.

A small passenger bus came slowly around the drive and stopped. Several men and women stepped out. They had been into town on the shopping bus. Away in the distance, against the boundary fence, a former bush nurse, Grace Perry, was hanging out her washing and talking to a group of Aborigines across the fence.

Down the pathway, two elderly people were pushed in wheelchairs by members of the staff. The branches of the big old gum trees hung over them in protective care. Away in the distance, the Macdonnell Ranges carved their character across the horizon.

'Come in and have a beer when you're finished,' called out old Bob Pasternack.

Gisele grinned at me.

'You can knock off work for the day,' she said.

14

The Defiant Hope

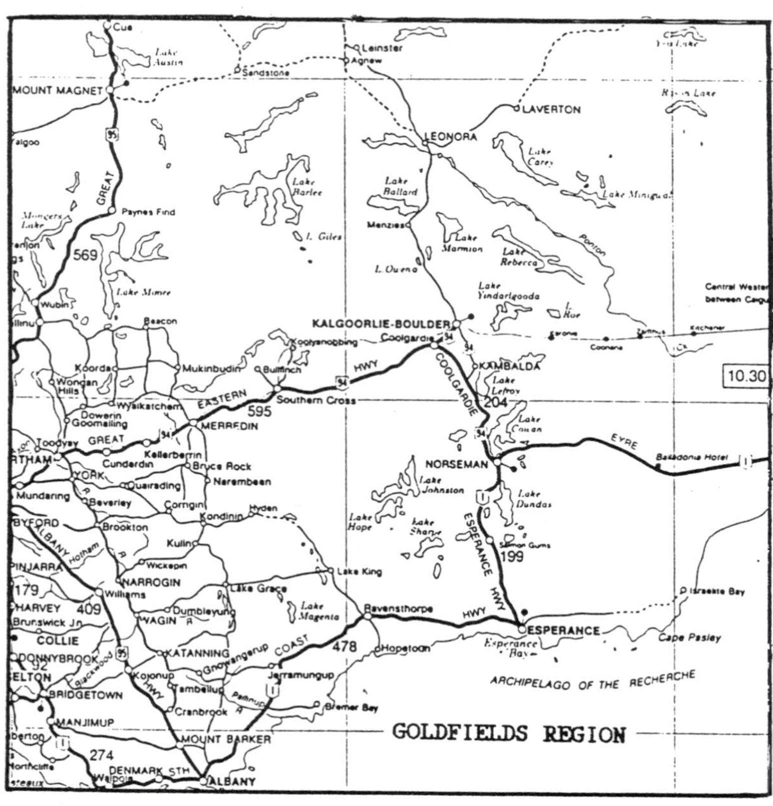

The young policeman turned the vehicle off the sand track and pulled up. 'Here's the dam,' he said cheerfully. 'We'll get out and have a look.'

A groan of dismay came from the back seat of the four-wheel drive.

'Do we have to?' asked one of the nurses, reluctantly. 'It's probably empty and I'm not going to climb that bank just to look at a lot of dry mud.'

I climbed down from the vehicle and stretched. The hot sun beat down mercilessly. The prospect of scrambling up the pile of yellow clay that was the wall of the dam was not all that exciting, especially if the dam turned out to be empty.

But this was no ordinary dam. This was the Olympic Dam in South Australia, after which the big uranium, copper, mining project had been named. People were sure to ask if I had seen it, as they do with any odd place in the outback. So I set off after the young policeman, as he began to climb. I had just started when I heard the door of the vehicle slam and the voices of two disgruntled nurses indicated that they were not far behind.

We reached the top of the bank and looked down. There was more than mud at the bottom. Thirst-crazed cattle had somehow scrambled to the top of the bank and then made their way to the bottom. It was a distressing sight. There were carcasses everywhere. Some had attempted to climb out again and simply collapsed and died on the way up. Others had blundered into the mud at the bottom and had been unable to get out.

'I don't want to stand here and look at it,' said one of the nurses.

She made to turn and head down again, when the other nurse suddenly pointed to the bottom and said, 'I think there's one of them still alive. See that head sticking out of the mud? I'm sure I saw it move.'

The fierce glare of the sun made it difficult to see clearly, but I could make out the head of a steer, just protruding from the mud.

'Well, it's not long for this world,' said the policeman, also turning to head back.

'You're not going to just leave it there?' said the nurse indignantly.

The young policeman looked a little sheepish.

'I think it's too far gone,' he replied. 'And even if we could get it out, it'd probably only last an hour or two.'

'Well, I think you are very callous,' said the nurse.

The young policeman looked at me for help. I had the feeling that he was a bit interested in the nurse and didn't want to jeopardise his chances.

'Let's go down and have a closer look,' I suggested, tactfully.

We slithered down the bank towards the bottom. The sight of the dead cattle was even more distressing at close range. We walked along to the nearest point where we could inspect the steer. It was a metre from the edge and only its head and front shoulders remained above the sickly yellow mud.

'It's alive all right,' said the young policeman. 'But it's too far down in the mud to get out.'

'But look at the poor thing,' said the second nurse, who had decided to take up the cause of justice. 'Look at its eyes. It's appealing for help.'

The young policeman and I looked at each other helplessly.

'We might move it,' I suggested, 'if one of us stands here and pulls on the horns and the other one gets into the mud and pushes from behind.'

The look on the policeman's face suggested that if I were foolish enough to believe that, then I should be the one to get into the mud.

'What would John Flynn have done?' asked one of the nurses. She had a straight face, but a look of mischief in her eye.

This was one of the occasions when being the boss had its drawbacks. I took off my shoes and socks, rolled up my trousers and waded into the mud. It smelt horribly and I had the feeling that I would too, for some time to come.

The policeman took the bull by the horns and I pushed on its hindquarters. We heaved and struggled for a while without result. Then it appeared that the steer itself began to regain some strength. Eventually it began to move. I pushed harder, slipped in the mud and fell on my face. I got up, wiped the mud from my face and started to push again.

As the steer seemed to sniff the air of freedom, it began to thrash around in the mud and claw its way out. Finally, with a great effort from the three of us, the steer was out of the mud and lay panting on the ground alongside the policeman.

I stood knee deep in mud and looked at them.

'Now you can get me out,' I ordered.

The next job was to get the steer up the bank. We pushed and shoved and dragged the steer while the nurses stood at a safe distance, encouraging us. Finally we made it to the top of the bank. The steer was shaking with exhaustion and obviously very weak. The young policeman gave it a slap on the rump and it staggered down the slope. It slipped and fell a couple of times, but picked itself up and continued on.

The last I saw of it, the steer was stumbling into the distance, headed for goodness knows where.

The instinct for survival is strong in the bush, both for animals and human beings. It explains why people stay there, even when they are overwhelmed by disaster and ill fortune. Hidden within the instinct for survival, sometimes very deeply, is the element of hope. Even when the odds against them are impossible, the people of the outback never quite seem to give up hope.

Nowhere is this more clearly exemplified than in the Goldfields region of Western Australia.

'Hello, Max. What are you doing down this way?'

I looked up from the paper I was reading and saw the bluff, genial face of Sir Charles Court looking down at me. Sir Charles was, at that time, Premier of Western Australia. We were both on a plane headed for Kalgoorlie.

'I'm considering putting a patrol padre in the Goldfields region,' I replied. 'The Pilbara and the Kimberley have had all the attention for the past decade. I reckon the Goldfields have been neglected for too long.'

The Premier's face lit up.

'I'm going down to try and put some heart into the people of Kalgoorlie,' he said. 'They've had a pretty rotten run lately. Do you mind if I quote you in my speech? I don't have a lot to offer them and it would be nice to say that someone else is interested in their welfare.'

The Goldfields had been a glamour region of Australia in the '60s, just as it had in the gold-rush days, a century before. The new 'gold' of the 1960s was nickel. A whole economic boom was built on the discovery of nickel and the prospects of its mining. Ordinary people all over Australia became players on the stockmarket, and names of companies such as Poseidon were household words.

Then the bubble burst. Share prices tumbled and hit the stock exchange floor with a crash. People wrote off their losses and charged them up to experience. But the people of Kalgoorlie and the surrounding region were hit heavily. When the dam of prosperity dried up, the Goldfields region lay at the bottom, up to its neck in mud and still sinking.

The plane on which I was travelling landed at Kalgoorlie. Sir Charles wished me luck in the new venture and I wished him luck in his efforts. He needed it more than I did.

'I've booked you into a motel,' said Ken, who had met Doug and me.

As we drove through the streets of Kalgoorlie, he pointed out buildings which had been erected in anticipation of the new mining boom.

'They'll never be used now,' he observed.

We arrived at the motel. It seemed to stretch for miles.

'This is another casualty,' said Ken. 'It's supposed to be the biggest in Australia. They built it in anticipation of thousands of people coming with the boom. I've no idea what they will do with it now.'

We settled in our room and yarned with Ken for an hour or so.

'I'm feeling hungry,' said Doug. 'What say we order some sandwiches for lunch.'

I picked up the phone and made the order. Then we continued to chat for a while.

'Those sandwiches are a long time in coming,' said Doug.

We waited for an hour. Then I got on the phone and made a polite enquiry whether the bread was baked yet. The voice at the other end was profuse in apology.

'Dreadfully sorry, sir. The girl who is bringing your lunch has made three attempts to reach you and has got lost every time.'

Later we drove around Kalgoorlie. The centre of the city was most attractive. Some of the old hotels had been built in the grand fashion of the golden years and had been beautifully restored. Not so the churches. They bore the brunt of neglect. Many had been closed and their emptiness gave them the appearance of hollow sepulchres.

'This is Hay Street,' Ken told us. 'I thought you'd want to see that.'

Hay Street was the site of a row of brothels. It had a place in Australia's folklore, similar to that of Ned Kelly. In the clear light of a Kalgoorlie day, the tawdry little houses looked pathetic.

As we drove away from the city centre, the weariness of age and the essential ugliness of the old goldmine operations began to take a firm grip on the landscape. The head-frames of the mine shafts stood out like rigid skeletons on the horizons. The buildings were more scattered and dilapidated. Piles of tailings looked like giant burial grounds.

Ken then drove out into the bush for a short distance. He stopped the car and we got out and walked into the scrub. We came to a circular tin-clad shed with an open space in the middle. The atmosphere was one of drab emptiness.

'I thought you might like to see the two-up school,' he said. 'It's all part of the local culture, you know.'

Later, we drove through Boulder, the twin city of Kalgoorlie, which once had rivalled it in wealth and grandeur. Now it looked like an untidy cemetery on the outskirts of Kalgoorlie.

Gold had once been the lifeblood which pumped vibrant life into this region. As we drove back to Kalgoorlie, I felt that not even a healthy transfusion could save it. Perhaps the best Kalgoorlie could hope for was to be polished up like antique furniture and sold to tourism.

'You're staying with the Cleverleys tonight,' said Ken. 'You'll like them. Bill is a real identity in Kalgoorlie.'

We pulled up in front of a modest, tin-roofed cottage with a fresh paint appearance and a beautifully kept garden. The couple who greeted me were grey-haired and bright eyed, with the open healthy faces of country people.

'They tell me you're quite famous,' I said to Bill, over dinner.

'It comes of doing your job,' he replied, looking me straight in the eye. Then he told me the story. He didn't assume any air of false modesty.

'I was head of the Geology School at the Kalgoorlie School of Mines,' he began.

I had heard of the Kalgoorlie School of Mines. It was one of the great institutions which had nurtured the men who had gone on to carve their names in the history of mining.

'In my spare time, I used to help a lot of the prospectors who fossicked around the region. They would bring in rocks which they thought might contain ore. Around the '50s, when the uranium boom was on, there were a lot of fellows fossicking in their spare time.' Bill laughed. 'Of course, most of the samples were useless, but I would analyse them and I didn't charge anything.'

Then one day, a fellow called George Cowcill came in. I knew him. He was a farmer who did a bit of gold prospecting in his spare time. But he got bitten by the uranium bug. And he brought in a sample which he thought had some uranium oxide. Well I analysed it for him and there wasn't.'

Bill paused reflectively.

'However, I like to do my job properly and I tested the sample for other traces. I discovered it had traces of nickel, very encouraging. I told George and suggested that he bring back some more. He did and there was certainly nickel there.'

'So what did George do?' I asked.

'He didn't do anything then. Gold and uranium were all the go and that's what he was after.' Bill Cleverley paused again. 'Ten years went by and I had forgotten all about the incident. But apparently, one day, George was drinking in a pub with a bloke called John Morgan. He was with Western Mining. And George happened to mention the nickel traces to Morgan. Well you know what those Western Mining fellows are like. One sniff and they're off like bloodhounds.'

'And the rest is history,' I commented.

'Yes,' Bill said, grinning. 'The place where George Cowcill found those nickel traces is now the Kambalda operation and what a blessing that's been to the region.'

I looked at this modest unassuming man, now retired and living in his modest unassuming home. What was it that he had said? 'It comes of doing your job.'

The meeting with Bill Cleverley was a good introduction to the Goldfields region, and especially to the mining industry.

Sometime later, we had established a padre in the region and I went to visit him in his home in Kambalda. It happened that his name was also Bill.

'The town's in a bit of a gloom at the moment,' he said.

'Someone died?' I asked.

'It might be the whole town, if the price of nickel doesn't rise. They're not replacing people who are leaving and there are gaps in the caravan park.'

We drove through the streets of West Kambalda, where he lived.

'Looks like the bloke who surveyed the town had a few too many when he laid out the roads. They wind around like agitated snakes.'

'It was a woman,' explained Bill, 'and before you make any sexist remark, she did it deliberately, to preserve the trees.'

As I looked beyond the town out towards the drab yellow desert country, I could appreciate the foresight of the planner's vision.

'The town seems quiet,' I commented. 'I suppose that's because of the work difficulty.'

'All the action's underground. We'll go down a mine this afternoon.'

'You have to take off all your clothes if you're going underground,' said the miner who was to be our guide. 'Underclothes, the lot.'

We stripped down to the skin and put on the clothes offered, including overalls, heavy boots and the mandatory miner's helmet. The guide lead us to a vehicle which was harshly functional and totally devoid of those optional extras so dear to the heart of car salesmen.

The entry to the Otter nickel mine at Kambalda was like a large concrete drain pipe sloping down into the bowels of the earth. We were immediately enveloped in the roar of the vehicle's engine. The sound reverberated against the hard walls of the tunnel and multiplied to a fierce degree.

The tunnel turned and twisted on its downwards path. I had the feeling that it was literally hell bent, while the driver acted with all the verve of a Grand Prix specialist as he guided us around the twists and turns. Then, as if guided by some sixth sense, he pulled over into a small recess in the side of the tunnel wall.

We sat in the dark and waited. Around the corner, steaming up the incline, came a huge metal monster. With menacing roar and blazing headlights, it forced its way past us and continued up the incline. The heavy load of ore it was carrying could not frustrate its intention to reach the top in record time.

In a moment it was gone and the silence returned. Our driver seemed in no hurry to return to his downward flight. It was just as well. Another metal monster suddenly appeared from above us. This time it was empty. It came down the incline at full speed, rounded the corner and disappeared on its downward journey.

We resumed our own plunge. From time to time we pulled into the side of the tunnel, to allow further ore-carrying trucks to pass.

At various depths on the journey, there were horizontal drives which led off the main incline. We turned off into one of them and travelled along a labyrinth of branches leading in several directions. There were no street signs to indicate where each of them was going. Without the driver, we would quickly be lost.

With no sky and no horizon, the world of the underground mine is a different world. It is a world which only those who work in it really understand.

The working space of the mine was severely restricted. There was no flexible space, as there can be on the surface. Tunnelling is an expensive business and no more tunnels are carved out than necessary. Yet despite this, there was an amazing amount of machinery operating.

In addition to the vehicles which carried the ore to the surface, there was machinery for lifting and dropping it to different levels. Much of the maintenance work was also carried out underground. It saved the time which would be necessary to get the equipment up to the surface.

The sense of being in a different world was reinforced by the clothing worn by the miners. Men moved about clad in what was basically clumsy apparel. They wore heavy boots and the traditional miner's hat with its lamp. Most also wore big gloves and protective glasses.

The general atmosphere was dark and dank. The hollow caverns of the tunnels were fitfully lit by a few lights, strung along the walls. When we got out of the vehicle, I picked my way along the tunnel, stumbling over pieces of rock and generally uncertain about what was going on.

By contrast, the miners seemed relaxed and at home. They greeted each other cheerfully as they appeared and disappeared into the enveloping darkness. The place belonged to them. They had a job to do. Although we were treated with courtesy, we were obviously an additional responsibility which could be tolerated for a short time only.

We came to the face of the tunnel where the mining was taking place. In contrast to the general gloom, it was lit with the harsh glare of floodlights. I expected to stumble across a television crew filming the operation. Large drilling machines attacked the whole face of the tunnel. Dust and steam created an eerie mist, made eerier by the glare of the spotlights. We were advised not to go too close.

Travelling through the mine was a bit like going through the river caves of a fun park. There were long stretches of darkness. Then suddenly we came to a brightly, but garishly lit scene. Then we were back in the darkness again, wondering what the next scene would be, or when we would return to the daylight again.

'We're working at different levels,' explained our guide. 'When the ore is drilled out, it is dropped down a shaft to the crusher, which is at a lower level. Then the trucks take it out.'

We were driven to the various levels. Unlike buildings above the ground, I had no sense as to whether we were 500 or 5000 metres underground. Mines are not places for people suffering from claustrophobia.

We climbed a ladder which had been hung to the side of a shaft. At the top was a narrow bench which had been cut into the wall. We inched along the bench, jumped over a gap and climbed over a heap of rubble. Above us, at the top of this heap, was the entrance to a hole which was just about wide enough for someone to crawl into. Somewhere up that inclined little tunnel, a miner was drilling out the ore.

The drill he used was very heavy and the incline steep. I found it hard to imagine how he had got himself up the shaft, let along drag a drill with him. It was a cramped, breathless and dusty atmosphere. The work was obviously physically hard. I had heard of the high wages which the drillers earned. Now I knew why.

I suppose the philosophy of this kind of work is to get in, get the ore and get out as quickly as possible. I could understand why.

We left the miners to their work and returned to the truck which would take us back to the surface. We had hardly started our journey back up the incline when the driver turned the vehicle into one of the recesses in the wall and stopped. This time the cause was not the impending descent of a monstrous ore carrier. We had a flat tyre.

There is something slightly absurd about having a flat tyre 1500 metres underground. To make matters worse, the driver did not have a key to the lock which kept the spare tyre safe. Some unorthodox mechanical surgery solved the problem. We returned to the surface, the fresh air and the clear blue sky.

'Those fellows down there deserve all they get,' I remarked to Bill. 'I wouldn't like to spend the rest of my life stuck up in one of those drives.'

'The money's good while it lasts. But the strain gets to some of them and to their families after a while. At times like this, the uncertainty grips the whole town. Everybody talks about it and they feel the isolation more than at any other time.'

The next day we left Kambalda and drove north to another part of the Goldfields region.

'This is where you really understand what the Goldfields is all about,' Bill told me. 'It's a very complex region; Aborigines, pastoralists and miners all mixed up together.'

We were driving through a little town called Menzies, about 300 kilometres north of Kalgoorlie. It was a relic of the gold-rush days. The sandstone buildings and cottages seemed to have changed little, despite a surge of new mining developments.

'The whole region is dotted with places like Menzies,' commented Bill. 'Leonora, Sandstone, Cue are just a few of the little towns that refused to die. They look like ghost towns, but there are people who live in them. Sometimes I wonder why.'

North of Leonora we came to the functional no-nonsense town of Leinster, which serves the Agnew mine.

'How do the pastoralists get on with the miners?' I asked Bill.

'Leinster's been good for the station people up here,' replied Bill. 'It's got a medical centre, a school and stores with a reasonable supply of goods. But perhaps the most important thing is the commuter plane that flies regularly to Perth. That's made a hell of a difference.'

'Anyway, you can ask the people themsleves. We're going to drop in at one of the stations.'

'Cattle?' I asked innocently.

'Sheep.'

I looked across the broad sweeps of relentless red soil, without a trace of grass. There were no rivers or creeks with billabongs and gum trees under which the sheep could shelter from the scorching heat of the sun.

'Sheep survive up here and so do we.'

The speaker was Don, who owned the station we were visiting. We had gone out on a bore run, inspecting the windmills and water troughs which were located all over the property.

'But it's a matter of eternal vigilance,' he continued. 'The bore run is probably the hardest and most important job I do, especially in summer. We have to ensure that the windmills are working and the water troughs are full. If we don't do it twice a day in the summertime, we stand to lose hundreds of sheep.

'It's a time-consuming, monotonous and tiring job. In the height of the summer especially, doing the bore run twice a day is exhausting. But it's the key to survival, for the sheep. And that means survival for me.'

'What kind of thing can go wrong?' I asked.

'There are two things the pastoralist looks for on the bore run. One is a windmill which is not pumping the water. The other is a blockage in a water trough, which is preventing it from filling. This can be caused by an animal, crazed with thirst, scrambling up into the trough and dying. Pastoralists yearn for the day when it will be possible for the supply of water to be guaranteed, without the time-consuming and enervating task of the bore run.'

'We thought we had the answer,' Bill chimed in. 'I was up here visiting Don and went out on the bore run with him. The temperature was particularly high as we did the rounds fixing a few problems along the way. Eventually we returned to the homestead and slumped down in a couple of armchairs for a beer.'

'Then,' said Don, taking up the story, 'Bill uttered those immortal words: "There must be an easier way of doing this".

'We figured out that if the windmills were not working, then the troughs could not be filled. If the windmills were working, but something was blocking the troughs, then the same result would apply. So the key to solving the problem was to find a way to monitor the water level in the troughs.'

Bill and Don hit on a scheme of floating a device in the trough, which would set off an alarm when the water dropped below a certain level. That alarm would be transmitted by a radio beam to the homestead. If the station manager read the receiving machine in the homestead twice a day, he would know when and where there was trouble. The transmitting devices at each windmill would be operated by solar power. The sun was the one thing they could be sure would work.

'How did you get on?' I asked.

'We took the scheme to a leading electronics manufacturer. They were interested and after they tested it, believed it would work.' Don grinned. 'There was one funny side-issue in the process of convincing them.

'One of their engineers believed that the device would work better if it were attached to the ballcock, which floated in the trough, and acts in controlling the water level, as it does in a cistern. I told them their idea wouldn't work as the kangaroos, who also need a drink, learned to depress the ballcock with their paws, to bring up the level of the water.

'The American engineer didn't believe him,' said Bill. 'So Don took him out on a bore run. They waited near a water trough. To the astonishment of the American, a kangaroo hopped up to the trough, depressed the ballcock with his paw, and calmly drank the water.'

Bill and Don were very confident that they had solved the problem. Unfortunately, the company which was interested in the device came to the conclusion that the number of the water warning devices which they might sell did not justify the investment of the capital required to produce it. It is a sad comment that a device which would have been so beneficial to the pastoralists of the outback and elsewhere could not be produced because of financial constraints.

Getting down to business. Prime Minister Malcolm Fraser meets with Aboriginal leaders on the Todd River bed, Alice Springs.

Scene of a small victory. Aboriginal fringe camp at Mount Blatherskite, Alice Springs, where houses were later built.

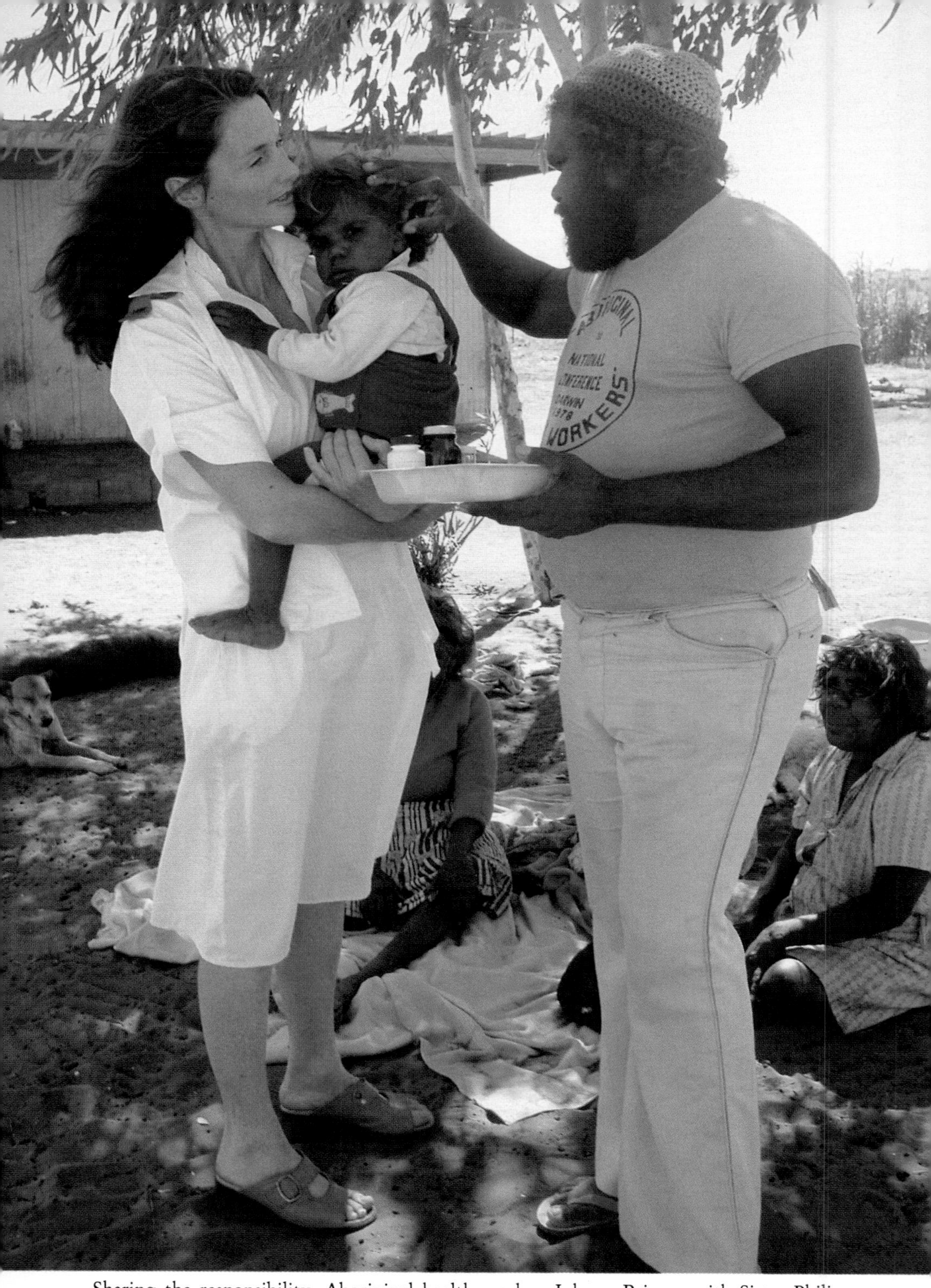

Sharing the responsibility. Aboriginal health worker, Johnny Briscoe with Sister Philippa Rhodes at Apatula Community, Finke.

'Well, since you're here, you might as well jump down and help.'

The voice which came floating up from the bottom of the water tank, belonged to the manager of the next pastoral station we visited. We had poked our heads over the edge of the tank to say hello. The tank was empty, except for the sludge at the bottom. The manager was up to his knees in it, cleaning it out. We took off our boots and socks and hopped in to help.

An empty water tank is a symbol of a bad drought. It is the ultimate symbol of despair for the pastoralists. Fires can be fought and floods ultimately recede. But facing a drought which might go on for years can drain all your energy, determination and faith.

I asked the pastoralist why he bothered to clean out the tank.

'I suppose it's a symbol of faith. If we don't believe the rains will come, we may as well move out.' He looked at me and straightened his shoulders. 'So I'm getting ready for the rains.'

In the loneliness of the outback, with no-one to talk to, the only way to cope is to keep on doing things that at the time may seem trivial and futile. But they are acts of faith.

As droughts are times of desperation for the pastoralists, so a collapse in metal prices brings times of desperation for the people of outback mining communities. Anxiety gathers over the town like a black cloud. News and rumours of further falls in prices circulate among the workers and their families. Perhaps one or two people in management are transferred to other locations. Overtime begins to contract.

Empty spaces begin to appear in the caravan parks. That means the independent contractors are leaving. Some of the company houses become vacant. Everyone is talking about what might happen next. When will the retrenchments begin and how many will lose their jobs? There is less spending in the shops and fewer trips to the capital cities.

When both a drought and a collapse in metal prices coincide, as they have done in the Goldfields region, then depression really sets in. The harsh fact of life is that those most affected by the drought or the falling metal prices can do nothing. Except, perhaps, get up and go somewhere else. In that respect, leaving is like death. For both pastoralists and miners, it is more than simply losing a job. It is the death of a way of life. To go and live in a major city means beginning life again in a strange, unaccustomed world.

What the people experience at a time like this is a kind of grieving, and these are the times when the work of a patrol padre like Bill is most important. These are the times when, after a few media headlines about falling prices or long periods of drought, regions such as the Goldfields once again become the forgotten places of the nation.

That was why we decided to appoint Bill to the Goldfields region. It was not an earth-shattering decision. It was not likely to bring rain or raise metal prices. But perhaps it helped to demonstrate that sometimes the Christian Church means what it says it believes.

Bill told me how he worked hard to organise a Christmas service during the drought.

'A family offered to hold it on their property. I sent out invitations and as far as possible, visited families and urged them to come. The attendance was very good. We had a Christmas party afterwards and it was a great occasion.

'Shortly after the service, the rain fairly belted down. What happened then has become a local legend. People swore that those who attended the Christmas service received a huge amount of rain. Those who didn't attend only received a little.'

Bill laughed out loud.

'I only wished I had another service the next Sunday. I think I could have cashed in heavily.'

Sometimes in the Goldfields region, as in other parts of the outback, the people wait for years for the rains to come. While they wait, things can happen which refresh the dry bones of their existence and remind them that hope is never lost.

Occasions such as the Christmas service which Bill conducted might be rare. But you will find in the outback that people talk about them and remember them for years to come.

John Flynn went into the outback at the beginning of the century and discovered a people who were fighting a grim battle for survival. He also discovered that, hidden within that struggle, was the tiny flame of hope. Sometimes it seemed to be all but extinguished. But Flynn brought to the outback medical and spiritual services which nurtured the flame of hope and made it grow stronger and brighter.

Australia is still a frontier country. Droughts, bushfires and floods still afflict its people. Falling metal prices and other factors still affect those who work in the mining industry. Aboriginal health problems and their struggles for dignity and justice are still acute and painful.

But the flame of hope still burns, because the people of the Australian outback are a very special breed.

15

The Hungry Heart

'They tell me you're leaving.'

I looked across at the couple with whom I was sitting on their homestead verandah. I had sat there often and enjoyed their hospitality and company. They looked back at me with a mixture of sadness and resolution and perhaps a touch of guilt. They had obviously made up their minds to leave the station, but the decision was costing them a lot of anguish.

'We just feel that the children deserve a good start in life,' explained the man. 'Of course, we could send them to boarding schools. I went to one myself for six years,' he added with some pride.

'But things are different now,' said his wife. 'Children need all the help they can get with their education. They talk about providing bush children with telephone hook-ups and satellite television and it all sounds wonderful. But you need libraries and science laboratories and computers and most of all, the face-to-face things, not only with the teachers but with other children as well.'

'We could send them away to boarding school and they would get all that,' the husband continued. 'But the more complicated it gets, life I mean, not just education, the more we feel we should be with them. So we think it's time to go.'

He paused for a moment and looked out across the open country. His mind was made up, but his heart was still rebelling.

'I think the adjustment will be harder for us than the kids,' he added in a low voice.

The situation of this family was one of the many cases in the outback where, in the end, isolation was the winner. It was said of Flynn of the Inland that he made family life possible in the outback, through his medical mantle of safety. But with the passing of the years and in common with Australians everywhere, the people of the outback began to expect more for their children than just the assurance of survival. The hungry heart is always looking for deeper satisfaction.

Even some of the old-timers began to retire to the south.

I was visiting one of the well-known and long-established mining towns in northern Australia and was looking forward to calling in and having a chat with an old mate.

'He's gone,' said a mutual friend, when asked.

'Bill gone?' I replied in astonishment. 'You mean he's dead?'

'No, he and Joan have gone south. Victor Harbour I think it is.'

We drove slowly down the main street of the mining town and pulled up outside the little iron-clad cottage where Bill and Joan had lived for decades. I closed my eyes and recalled the neat old-fashioned simplicity inside. When I went through the front door of that cottage, it was like stepping into a past era.

'How will they survive down there?' I asked.

'Well,' my friend replied slowly, 'Their son and his wife and the grandchildren are down in Adelaide and that's always a big pull. Besides, Bill and Joan are not getting any younger. He's nearly eighty you know.'

He paused for a moment and squinted his eyes as he looked through the harsh glare of the sun down the wide main street. The bitumen was almost at boiling point.

'Then there's Joan's health. She made three trips to Adelaide last year for treatment and it's a hell of a long way there and back, as well as the expense.'

I was a bit bewildered by all this. Bill and Joan were, in a way, symbols of the battlers of the bush who had invested so much of their life there that the thought of leaving was incredible. I remembered another old friend who had said to me once, 'The only way you'll get me out of this place, is to carry me out in a box'.

The radical change which happened in the Australian outback in the '60s and '70s brought into sharper focus the isolation in which so many people lived. But it also brought into question the need for people to live in that isolation.

Nowhere has this been more apparent than in the mining industry.

'We've got a new mining project opening up in the Kimberley,' said the mining executive.

'Anything for us to do?' asked Gray hopefully.

'You'll have to excuse my colleague,' I explained. 'He spent eight years in the Pilbara and he's never happy unless he's working with miners.'

'Oh, we wouldn't trust you fellows with this one,' said the executive, with a sly grin. 'It's a diamond mine.'

'Where are the diamonds located?' I asked. 'Or is that a state secret?'

'It's at the bottom end of Lake Argyle. They've discovered a kimberlite deposit, like a hard perpendicular core that goes down the centre of a hill. Actually, water has washed some of the diamonds down a gully into a creek bed, so there's some alluvial mining to be done first.'

'You mean that you can actually pick up diamonds out of the creek bed?'

'Don't rush up there and try,' warned the executive. 'The security is very tight.' Then he became serious. 'Actually, we do have a job for you. But, to be truthful, we are not sure what it is or when it will start.

'The big problem is where to locate the work force. We have three options. One is to build a town on site, like most of the other outback mines. The second is to use Kununurra, which is at the top end of Lake Argyle.'

'You could take the workers down the lake on hydrofoils, like they do on Sydney Harbour,' I suggested hopefully.

'What's the third option?' asked Gray.

'The third option is to fly the work force in and out of Perth.'

There was a long silence.

'But that's over 3000 kilometres. It'd be like having your workers in Perth and flying them to Adelaide.'

'Nevertheless,' replied the mining executive, 'it's the option we favour. We think in the long run it will be less expensive and create fewer problems. There's a growing opinion that the workers are happier when their families live in the capital cities and they can go back home.'

'Well, it certainly worked at Moomba,' I said after a while.

The work force at Moomba natural gas field, in the far north of South Australia, was entirely made up of commuters. They lived in dormitory accommodation at Moomba for a couple of weeks, and then returned south to their homes and families in Adelaide. It seemed to work well.

The final decision for the Argyle diamond mine was to fly the workers from Perth to the Kimberley for three-week shifts, but the executive acknowledged it wasn't going to be easy.

'There will be a construction force on site for about two years. Only a handful of them in the beginning, but about 2000 at its peak. Then it will taper off.

'What do you want us to do?' I asked.

'Well, it's a bit complicated.' The executive chose his words carefully. 'You know what construction workers are like. They want to bring their families with them. There's no town or settlement within a bull's roar of the place. So we'll have to make some provision for the women and kids, as well as the men.'

'Where will they live?' I asked.

'In caravans.'

I could see the problem. The prospect of providing for a community of 2000 or more, in one of the most remote parts of Australia, with a life expectancy of two years, was a planner's nightmare, especially when the numbers would vary enormously during that time.

'So what do you want us to do?'

'We want you to run a hospital for us and do something for the children.'

'You mean a preschool?'

'Yes, at least that, and maybe something for the older kids as well.'

'You mean, like a school?'

'Yes,' he finally acknowledged, 'like a school.'

'What about the Education Department?' I asked.

'They don't want to know about it.'

'I don't blame them,' said Gray. 'You don't know whether there will be six kids or 600 and how long they'll be staying. Still, compared with the problems you'll be facing with the whole construction work force, I don't suppose it's all that much. If Max is agreeable, we'll have a go.'

I was and we did.

We flew down to the mine site at a place called Limestone Creek. One of the managers took us over the area in a little helicopter.

'What sort of medical problems do you envisage?' I asked.

'A couple of weeks ago a fellow had a fall, out in the open country and broke his leg.' The manager grimaced. 'Road access was difficult, so we sent in a little machine like this.' He tapped the body of the helicopter. 'They strapped him onto one of the landing runners. The nurse put a pillow under his head for a bit of comfort and the pilot took off.'

He paused and then went on.

'What happened next was that the pillow slipped out from under the man's head. It got sucked up into the helicopter rotor. The chopper stalled and crashed to the ground. So we had three casualties to get out, the patient, the pilot and the nurse.'

When the construction eventually started, we opened a small nursing outpost hospital and a preschool centre. The Education Department had an attack of conscience and opened a small school.

The nurses and teachers who worked at Limestone Creek never knew what would happen next. One day we would receive a report saying that they were twiddling their thumbs. The next day there would be a frantic telephone call demanding two more nurses or another teacher. Unfortunately, the outback doesn't have its own version of Rent-A-Nurse.

We commenced work at the Argyle diamond mine in April 1984 and finished in November 1985, a period of about eighteen months.

Seventy years earlier, in 1915, John Flynn had begun to dream of building a hospital in Alice Springs. It took him ten years to fulfil that dream and another twenty before the need to operate the hospital came to an end. The outback had come a long way since then.

Not everyone in the outback has the freedom to choose where he or she will live, nor would everyone necessarily choose to leave the outback. So the battle to overcome isolation still continues.

One of our travelling nurses in the far north-west of Queensland once asked to be made a justice of the peace. When I asked why, she replied, 'Some of the families I visit live over 500 kilometres from Mount Isa. Would you like to travel 500 kilometres, just to have a statement witnessed by a JP?'

The people most affected by the changes of the '60s and '70s were the Aborigines. Removing the isolation in which they lived was not always welcome nor helpful.

My work brought me into constant contact with Aboriginal communities. They often expressed appreciation for the work we did and I made many friends. But there was an underlying feeling that we were strangers, who, however friendly and supportive, did not really belong. The heart of the Aborigine hungers for things we could not give.

'I reckon we're coming in a bit high,' said Gordon.

I looked at Carold, who was flying the plane. His face was a study in anxious concentration.

'It's the hot air,' he told us. 'It's making us float and the plane won't drop fast enough.'

'Well, I think we might run out of airstrip,' commented Gordon casually. 'See that little track that runs away from the end of the airstrip?'

Carold nodded.

'I think it's wide enough to take our wheels. So just keep going.'

The Cessna 182 ran off the end of the strip and sped along the twin wheel tracks. Carold eased back the throttle.

'Don't stop now,' said Gordon. 'The track will take us into town.'

Gordon was not one for walking too far in the hot sun of central Australia, and Carold taxied along until the tracks ran out. Just ahead of us was a scattering of bush buildings. We came to a halt and killed the engine. I looked at the little settlement.

'So this is Finke,' I said.

People who used to travel on the old Ghan railway line from Port Augusta to Alice Springs will remember stopping at a railway siding with a large sign, bearing that name. A few wooden cottages near the siding housed the fettlers who looked after the long stretches of railway line that crossed the desert.

The settlement was named after the age-old Finke River near which it is built. There will be some people who remember Finke more vividly than others. They were the unfortunate victims of the flooding of the Finke, which on the rare occasions it has happened, washed away the railway line and left passengers stranded for weeks at a time. The railways workers took the line of least resistance. They waited until the river had subsided and then simply threw another line across the river's sandy bed.

Like similar sidings on the Ghan railway line, Finke had become a gathering point for a small community of Aborigines. It also served the surrounding cattle stations, such as New Crown and Andado.

Gordon and Carold and I had flown into Finke from Oodnadatta, further south. Apart from the problems in landing, the trip had not been difficult. We just followed the railway line up the centre of Australia.

We got out of the plane and stood in the strong clear sunshine. It was the middle of the day and there was no-one in sight.

'Well,' remarked Gordon, 'the Aborigines don't seem all that impressed that the white gods have dropped out of the sky. We'd better try and find someone.'

The roads were unmade and the orange sand crunched under our boots. A low stone building with a wide verandah bore a sign which announced that it was the Finke Hotel. Two Aborigines sat on the ground under the shade of the verandah.

'Where does Margaret live?' I asked politely.

'In caravan,' replied one of the Aborigines, twisting his head to point the direction.

We turned and set off down the road.

There is something a little absurd about walking up to a caravan in the middle of the desert and knocking on the door. But there was nothing absurd about the woman who opened it and invited us in. Margaret was a tall woman with straight silver hair. Her tanned skin had not been obtained from systematic sunbathing, but from twenty-two years living with Aboriginal people in Central Australia.

'We want you to open a medical clinic in Finke,' she said. 'We have about a hundred Aboriginal people here as well as the whites who work on the railway. There's no-one with any medical training.'

'So what happens in an emergency?' asked Gordon.

'The flying doctor comes down from Alice,' explained Margaret. 'But he's not always available. We had a serious case a few weeks ago. There was a gang of railway workers about 100 kilometres down the track. One of them put an axe through his foot. He nearly severed it. They brought him back here as quickly as they could, but it was a long journey.'

'What happened to him?' I asked.

'The flying doctor plane was out somewhere,' said Margaret. 'So we had to drive him to Alice Springs. It's about 300 kilometres.

'We poured Dettol on his foot, wrapped it in a towel, and put him on the back of a truck. We didn't have any anaesthetic, so we gave him a bottle of whisky. He survived the trip and so did his foot. But no-one was exactly thrilled with the experience, especially the man.'

'How about the Aboriginal people?' I asked. 'What is their health like?'

'The same as Aboriginal people everywhere. We desperately need someone to introduce health programmes, especially for the children. The community is doing some good things. They are building their own houses. But health is a major problem.'

We established a medical clinic at Finke, appointed two nurses and worked there for five years. It was an extraordinary experience.

'This is Johnny,' said Philippa, one of the nurses. 'He's our new health assistant.'

I saw a round-faced, round-bodied, smiling Aborigine. He looked up from the Aboriginal child he was treating.

'Johnny's now doing basic treatments like cuts and infections,' explained Philippa. 'He also goes over to the school every day and gives the children their tablets and inspects their eyes and ears. He is learning very quickly.'

Just how quickly Johnny learned was a matter of amazement, not only to the nurses, but to the flying doctor who came regularly to conduct clinics.

'He can't read or write, you know,' said the flying doctor to me on one of my visits. 'But he seems to know exactly what medicines to give. He must have a fantastic visual memory.'

About twelve months later, I received a letter from the nurses at Finke. 'We feel the time has come for one of us to leave,' it said. 'Johnny is handling the work very well. But the only way he can advance further is for one of us to leave, so he will have greater responsibility.'

I visited Finke some time after one of the nurses had departed.

'How is Johnny coping?' I enquired.

'Very well,' replied the remaining nurse. 'I can now go out on a clinic run to the stations and stay overnight. That saves a lot of wear and tear on me. Then when Johnny is on night duty and the Aborigines come to him for trivial reasons, he sends them away with a flea in their ear. They take it from him. They wouldn't take it from us.'

'Neville, where can I get a dental chair?'

The big bluff country man looked at me as if I had taken leave of my senses. On the other hand, he was used to my outlandish requests.

'What do you want with a dental chair?' he asked suspiciously.

'It's for Johnny up at Finke. The dentist who comes down from Alice Springs occasionally is astonished at the speed at which Johnny learns. He's got him doing simple dental prosthetics and says that if we can get him a chair, he could do quite a lot of the treatment.'

'I know a dentist who changes over his chairs pretty often,' said Neville. 'It's probably a tax dodge, but I'll put the hard word on him.'

So in due course, a nearly new dental chair was dismantled in a town in southern New South Wales, packed into a light aircraft and flown to Finke. Visitors to the medical clinic of the little outback community were astonished to see Aboriginal patients reclining in the latest in dental chairs, while one of their own people skilfully filled or extracted their teeth.

The nurse who remained with Johnny spent more and more time away from the community. He grew in confidence and skill and attended special training courses held in Alice Springs. So I was not surprised when I received a letter from the Aboriginal elders at Finke. In essence, it said, 'We appreciate the help you have given us over the past five years. But we feel the time has come for us to take over responsibility for the hospital.'

It was time for us to go. We removed the remaining nurse from Finke and left Johnny in charge.

Six months later, I was flying around Central Australia with my bush mate Neville and a doctor called Warwick.

'Let's drop in and see Johnny,' I said. 'He passed on the word that he would like to see us if we were up this way.'

Since I had first flown up to Finke from Oodnadatta, the Ghan railway line had been moved 140 kilometres to the west. Finke was now a very isolated community. We landed on the airstrip. There was no-one to meet us. But once we got into the settlement, it was obvious that something was going on. There were people everywhere.

'It's the annual Alice to Finke motorbike race,' a passer-by told us. 'The first riders have come in. There's the usual crop of injuries.'

We hurried towards the hospital. Inside, there were a lot of people milling around. I looked for Johnny and found him. He was not his usual calm self, but he managed a welcoming smile.

'They didn't tell us what was going to happen,' he said. 'We weren't prepared.'

I discovered that no provision had been made by the organisers for a first-aid post. The old track down from Alice Springs was pretty rough and a number of motorbike riders had taken tumbles. One of them was serious. He happened to be a member of the police bike team and his superior was walking around the hospital, not looking very happy about the fact that the sole medical staff on duty was an Aborigine.

'Johnny,' I said, by way of introduction. 'This is my friend Warwick. He is a doctor and I'm sure that he would be happy to assist you if you wished.'

The police officer immediately turned to Warwick, assuming that he was now in charge. Warwick ignored him.

'If an extra pair of hands is any use,' he said to Johnny. 'I'd be happy to help.'

Johnny took him round and gave him an assessment of the various patients and what he had done for them. They came to the police rider who was obviously in some trouble.

'I think he got broken ribs and maybe punctured something,' said Johnny.

Warwick made a brief examination, then straightened up.

'Yes, well, I agree with your opinion. He should be taken to Alice Springs as soon as possible.' Warwick turned to the police officer. 'How about making yourself useful and getting a helicopter.'

The officer stood for a moment, open mouthed, and then hurried away.

Later in the day we flew out from Finke. I looked down on the tiny settlement. From the air, it seemed little different from the first time I had visited five years before. But a lot had happened and, despite the occasional drama, such as we had just observed with the bike race, the community was in far safer hands than when Margaret had first asked us to open a clinic.

Most importantly, the Aboriginal people were in command of their own destiny and, yes, it was time to go. Finke was one of the many projects in the outback, where the philosophy of John Flynn was realised, when he said, 'You succeed in this business when you do yourself out of a job.'

In the end, after eleven years, I had to face the question whether I had done myself out of a job. I guess it meant facing the truth about myself. I was not a bushie and never pretended to be and I was certainly not an Aborigine. I had come to love the outback and its people, but that is not sufficient reason for staying. To continue to battle against isolation in the outback requires that the source of satisfying the hunger of your heart is to be found there.

That is certainly true of most of the Aboriginal people and probably true of some white people. It is not the stuff of romantic yearning which has been the illusion that kept many white people in the outback long after it was time for them to go. It is a matter of facing the sometimes harsh reality of who you are and where your roots lie.

The outback had stimulated my life and given me wider horizons than ever I might have known or dreamed. But it was not the deep and continuing source of satisfaction for the hunger of my heart.

I was visiting the Kimberley region with a friend, named Harry. He was a keen fisherman and had brought along an impressive array of tackle.

At Fitzroy Crossing some Aboriginal friends took us out into the bush, to one of their favourite fishing spots on the river. The Aborigines and I contented ourselves with simple lines and used donkey meat as bait. We caught a few fish.

Harry, meantime, had taken his tackle further downstream and was standing in the water, performing some impressive casting. But he wasn't having any luck.

After a while, I gave up hanging onto a line. I walked up the steep bank and sat watching Harry weaving his luckless magic over the water. One of the Aborigines came up and sat alongside me.

'He no catch fish there,' he said.

'Why not?'

'Crocodiles.'

I scanned the water carefully. 'I can't see any crocodiles,' I remarked.

The Aborigine pointed to several spots on the river. As I watched carefully, an occasional ripple broke the surface. Eventually I was able to detect about six submerged crocodiles, floating in a semicircle around my unsuspecting friend.

'What are they waiting for?' I asked anxiously.

It appeared that Harry was standing near the place where the crocodiles came out and sunbathed on the bank. I hoped their patience and courtesy would last long enough for me to extract my friend.

I reflected that the sophisticated equipment and skills of our culture, and even dogged determination, were not enough to survive in the outback. You had to know it, respect it and belong to it.

The crocodiles were still biding their time as I slithered down the bank and began to wade into the water.

I came up alongside my friend and put an arm around his shoulder.

'Harry,' I said. 'It's time for us to go.'

Index